RIVERSIDE COMMUNITY COLLEGE
1916

D0762107

CENTRAL HEATING WITH
Wood and Coal

CENTRAL HEATING WITH
Wood and Coal

by Larry Gay

illustrations by Ken Rice

THE STEPHEN GREENE PRESS
Brattleboro, Vermont

To my faithful stokers: Carl and Jenny Gay

FIRST EDITION

Produced in the United States of America.
Designed by IRVING PERKINS ASSOCIATES.
Published by THE STEPHEN GREENE PRESS, FESSENDEN ROAD,
BRATTLEBORO, VERMONT 05301.

Library of Congress Cataloging in Publication Data

GAY, LARRY, 1937–
Central heating with wood and coal.

Includes index.
1. Dwellings—Heating and ventilation. 2. Fuelwood.
3. Coal. I. Title.
TH7466.W66G39 697'.03 80-29216
ISBN 0-8289-0419-7
ISBN 0-8289-0420-0 (pbk.)

Contents

Preface

This book is intended as general background reading for anyone contemplating installing a wood- or coal-fired furnace or boiler. It is not a catalog of furnaces and boilers currently available. Such a catalog would soon be out of date because of the rapidity with which new models are being introduced and older ones improved. Heavy emphasis has been placed on the details of how solid-fuel furnaces and boilers should be installed, especially in conjunction with the oil- and gas-fired units they supplement. Mistakes in installation are made more often than not; it is hoped that suggestions made in this book will not only forestall mistakes that undermine performance and safety, but also guide the reader's thinking on choosing a furnace or boiler, sizing the unit, and estimating the financial savings to be expected.

The book does not consider matters peripheral to furnaces and boilers, such as stoves, woodlot management, buying fuelwood, state or local codes pertaining to solid-fuel central heating systems. These topics are covered in hundreds of books and pamphlets available, often at no charge, from extension services and state energy offices. One peripheral issue has been included in detail—the production of domestic hot water from wood and coal—on the ground that there is very little authoritative information available on the subject. Since the payback period for a wood- or coal-fired domestic hot-water system is short—usually two to three years—these systems are easily justified economically. It is hoped that Chapter 5 will make the subject less formidable and speed the transition to wood-coal-solar hot-water systems. Solid-fuel and solar domestic hot-water systems are similar in many respects and complement one another nicely—sun doing the job in summer and solid fuel doing it in winter. Chapter 5 can be used to advantage by younger plumbers whose training has not included thermosyphon design, and also by owners of stoves who wish to capture some heat in the form of hot water.

Acknowledgments

Once again I want to express my indebtedness to Jay Shelton—friend and co-worker—for reading the manuscript with a sharp eye and detecting errors and weak spots. I also want to thank my partner, Louis Audette, for drawing my attention to unbearable turgidity in the more technical sections. Also, thanks are due the Country Journal for permission to use some material on furnaces that originally appeared there.

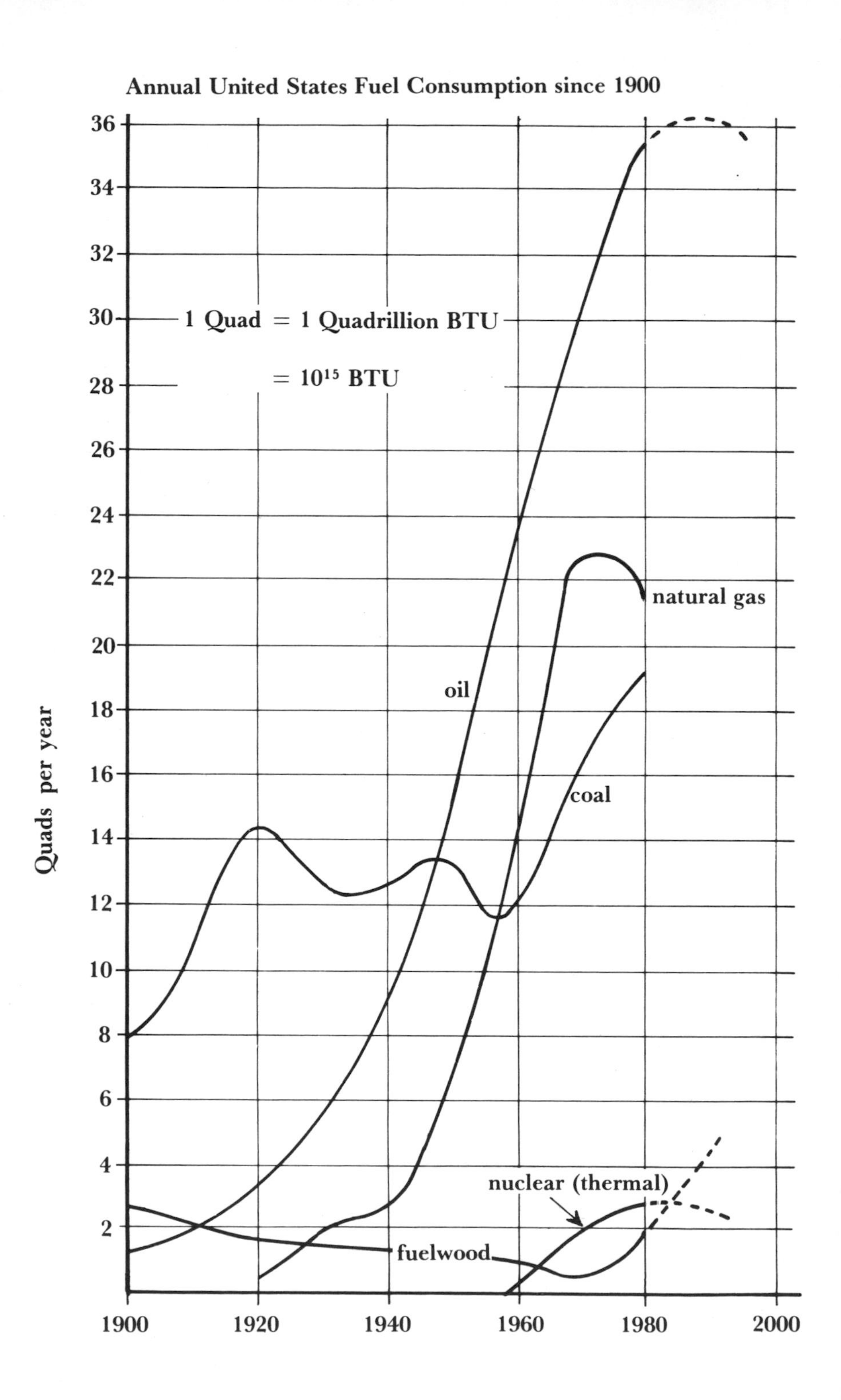

Annual United States Fuel Consumption since 1900
36
34
32
30
28
26
24
22
20
18
16
14
12
10
8
6
4
2
1 Quad = 1 Quadrillion BTU
= 10¹⁵ BTU
Quads per year
natural gas
oil
coal
nuclear (thermal)
fuelwood
1900
1920
1940
1960
1980
2000

Introduction

The figure on the opposite page shows the United States rates of consumption of our most common fuels in the twentieth century. The curve for oil has been a classic exponential with a doubling time of 15.2 years. That is, in March of 1915, twice as much oil was being used up annually as in January of 1900. By May of 1930, oil and gas were being used up four times faster than in 1900. Finally, in 1968, the exponential growth rate was broken. As we all know, that sort of thing can't go on forever. We are now (in 1980) at the peak of the oil curve. Soon we will be on the downhill side. This will be so regardless of political events in the Middle East or what Congress does or doesn't do. The fact of the matter is we are running out of oil, and it behooves us all to break our dependency on it. So far our record is not very good: In 1980, oil imports are actually above the level at the time of the first energy scare in 1973, in spite of a good deal of rhetoric. Fortunately they are now declining, but foreign oil burned in the United States still accounts for about forty percent of the total. This dependency on imports introduces instability into our political and economic system and has the potential to force some unpleasant changes upon us unless we get busy and do something about it.

There are vast resources of both wood and coal in North America. The wood resource is underutilized in the sense that the net annual growth rate exceeds the harvest by a ratio of something like four to one. Wood stoves and furnaces have come on strong since 1973, especially in rural New England, where oil prices are highest and the best hardwood in the world is readily available. By one estimate 1.2 million wood stoves and furnaces were sold in the United States in 1979. The growth curve for wood in 1980 looks like the growth curve for oil in 1910. However, wood energy will never reach the level of the oil peak. The total annual growth of wood in the United States is somewhere between ten and twenty *quads* (1 quad = 1 quadrillion = 10^{15} BTU). Four quads are harvested annually for lumber, paper and other wood products. Thus, somewhere between six and sixteen quads are potentially available for heating. Some of this potential fuelwood is in remote locations or on land owned by people who have no inclination to see it cut. Therefore, the wood energy curve will probably eventually peak well below ten quads annually. At two quads per year, heat from wood is about on a par with heat derived from nuclear fission in United States commercial reactors. The remarkable growth rate of fuelwood is proof that Americans can still spot a bargain. BTU for BTU, fuelwood is the cheapest fuel available in many parts of the United States today.

The energy potentially available from coal is far greater than that from wood, at least in the short run. By some estimates United States coal reserves amount to up to one-half the total for the entire world, although the actual figure is probably closer to one-quarter. Of this, only about one percent is clean-burning anthracite. Most coal mined today goes to industry and the production of electricity. Residential heating with coal has not come back from its peak, around 1920, as fast as wood-burning, partly because coal stoves and furnaces have not been readily available, and partly because the distribution network for residential coal has fallen apart in the last forty years. Even houses in Wilkes–Barre and Scranton, Pennsylvania, sitting on top of the best anthracite in the world, are heated by oil, gas, and now wood. The few companies making coal stoves and furnaces are expanding capacity as fast as they can to try to catch up with demand. The illustrations in this book reflect the fact that there are far more wood-burning appliances on the market than coal-burners.

Burning wood and coal is of course not without tradeoffs. Heating with wood or coal is always a lot of work. It can be frustrating. It entails some air pollution, not to mention ghastly chainsaw accidents and black lung. On the other hand, the alternative is staying hooked on oil, and that's not very attractive. There are those who heat with wood and coal quite successfully, thank you, and who smile every time the oil truck goes by. Our task must be to lower our expectations, to insulate, insulate, insulate, to organize ourselves and our machines to get the most out of every drop of oil, and to turn to solar heating as fast as we practically can. This will take time. Meanwhile wood and coal have an important role to play. I sincerely hope that this book will be a useful guide in taking a step that is not only to your economic advantage, but also in the national and international interest.

THE ORGANIZATION OF THIS BOOK

There is some information in the chapter on hot-air furnaces that is applicable to boilers. Chimney connections are the same in both cases, as are general comments about converting oil fireboxes to coal or wood. Thus, a reader primarily interested in a boiler may want to skim Chapters 1 and 2. Most drawings in Chapter 5 show a stove heating domestic water, but the drawings also apply to furnaces and boilers, since it is immaterial what kind of device heats the coil.

CHAPTER I

Solid-Fuel Furnaces

The reason my Uncle Jake had a coal furnace was so he could go stoke it. He was a faithful stoker, always coming up the stairs from the basement a little jollier than when he went down. You see, my Aunt Tillie ruled the house with a strict Puritanism, but her hip kept her out of the basement, so it was there that Jake took his nip. The furnace provided the excuse.

Other people have other reasons for having a coal or wood furnace in the basement: It saves money; it's more convenient than tending several stoves; it provides the evenly distributed heat we are accustomed to; it keeps dirty fuel and ashes out of living quarters.

With the price of oil going out of sight, a coal or wood furnace can save an impressive amount of money. **Table 1.1** shows the savings possible with a wood furnace costing $2,000 completely installed. The projection in Table 1.1 assumes no income tax credit, although tax credits are available in some states and may be available nationally in a year or two. Table 1.1 also assumes that the prices of oil and wood will continue to escalate at the rates at which they have escalated since 1973, namely at 7 and 3 percent, respectively, above the annual general inflation rate. This very uncertain guess is perhaps a little too optimistic, especially in the case of oil.[1] In any case, the investment in the wood furnace will be recaptured in two or three years, if one forgets to include a figure for operating and maintaining a wood furnace. An interesting way to look at the accumulated saving of $9,207 after ten years is to note that to realize the same payback on a conventional investment of $2,000, one

TABLE 1.1
Savings Schedule for a Typical Wood Furnace (1980 dollars)

	Price of Oil [a]	*$ saved on 1000 gal. oil*	*Price of Wood* [b]	*Cost of Wood* [c]	*Annual Fuel Saving*	*Accumulated Saving − $2,000*
80–81	$1.20/gal	$1,200	$75/cord	$468	$732	−$1,268
81–82	1.284	1,284	77.25	482	802	− 466
82–83	1.373	1,373	79.56	497	876	410
83–84	1.470	1,470	81.95	512	958	1,368
84–85	1.572	1,572	84.41	527	1,045	2,413
85–86	1.683	1,683	86.94	543	1,140	3,553
86–87	1.800	1,800	89.55	559	1,241	4,794
87–88	1.926	1,926	92.24	576	1,350	6,144
88–89	2.061	2,061	95.00	593	1,468	7,612
89–90	2.206	2,206	97.85	611	1,595	9,207

10,000 gallons of oil saved

[a] Escalation rate = 7%/year (general inflation has been subtracted out) for oil
[b] Escalation rate = 3%/year above general inflation for wood
[c] It is assumed here that 1 cord of good hardwood is equivalent in fuel value to 160 gallons oil.

TABLE 1.2
Heat Prices, March, 1980, New England

			Efficiency	*Price per useful therm*[c]
Oil	$1.05/gallon	= 75¢/therm[a]	65%	$1.15/therm
Wood [b]	$75/cord	= 27¢/therm	55%	.49/therm
Coal (anthracite)	$108/ton	= 40¢/therm	55%	.72/therm
Natural Gas	$7.25/1000 cubic ft	= 72¢/therm	65%	1.11/therm
Propane	75¢/gallon	= 77¢/therm	65%	1.18/therm
Electricity	6¢/kwh (kilowatt-hour)	= 175¢/therm	100%	1.75/therm

[a] 1 therm = 100,000 BTU
[b] 28×10^6 BTU/cord = 280 therms/cord (This is good hardwood, e.g., hard maple, yellow birch, white oak.)
[c] This column takes into account the efficiency of the heating device.

would have to get a return of 25 percent per year compounded annually after correcting for money's depreciation through general inflation. Most people will not be able to find a better way to invest their money today than to buy and use a wood-burning furnace.

The savings shown in Table 1.1 would of course be less in urban areas where wood is more expensive. Savings on a coal furnace would be less than on a wood-burner in most parts of the country, unless you happen to be close to a mine and can buy it for $73 per ton in 1980. **Table 1.2** compares the prices of various fuels as of March 1980. These may be wrong by as much as 50 percent for some areas because of special local conditions.

FURNACE ANATOMY

Besides a firebox, a furnace is composed of a heat exchanger, a blower to distribute hot air to the house, and, usually, automatic controls (**Figure 1.1**). There may or may not be a coil in the firebox or bonnet for heating domestic tap water. The jacket, of galvanized steel, and the ductwork make a furnace a furnace and not a stove.

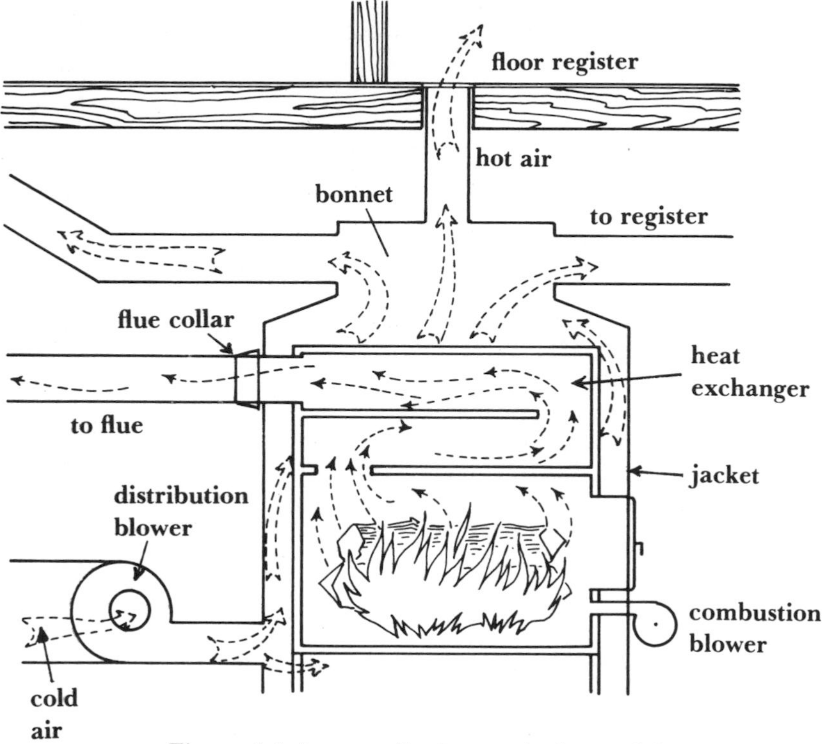

Figure 1.1 A generalized cross-draft wood furnace.

Or at any rate that's the distinguishing feature according to the Department of Taxes in Vermont, where the legislature has voted a tax credit of 25 percent of the installed cost of a wood furnace or boiler. Whether it's big or small, upstairs or down, manual or automatic is irrelevant to the revenuers in Vermont. Usually, however, furnaces are bigger than stoves, downstairs rather than up, and automatic rather than manual. True to form, the longer word with the French root labels the fancier item in English, and the one-syllable word of German origin is applied to the humbler heater.

The Heat Exchanger. A heat exchanger—a steel box of some kind between the firebox and flue—is characteristic of furnaces in contrast to most stoves. It provides plenty of surface area through which heat from the combustion gases on the inside flows out into the air stream that will eventually warm the upper parts of the house. The source of this air stream is the distribution blower, which takes cool air from the "cold"-air return and blows it past the firebox and past the heat exchanger. Warmed, the air stream returns to the house living area through hot-air ducts.

A heat exchanger (**Figure 1.2**) is needed on most solid-fuel furnaces because they are usually controlled the same way that oil furnaces are and not the way stoves are. That is, the fire is either on or

Figure 1.2 Cutaway view of the Gemini 5000 furnace showing heat exchanger. Hot gases from the fire flow through a compact array of tubes. House air is blown between the tubes.

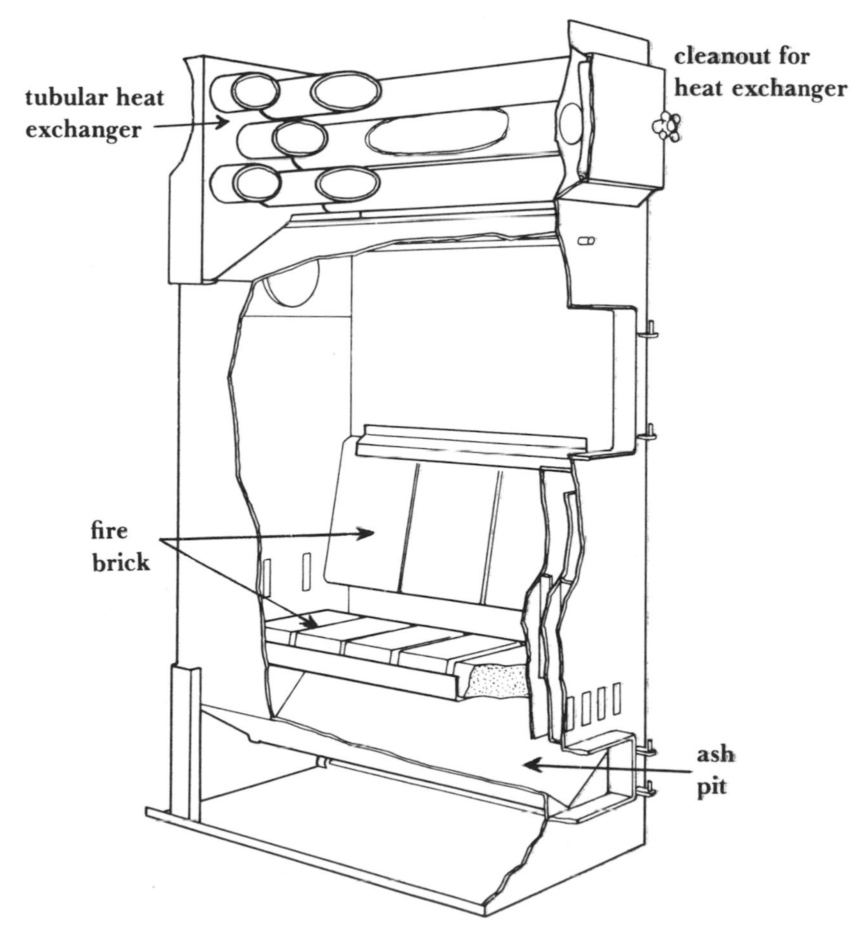

off. In fact, of course, in its "off" state the solid-fuel fire is not dead; nevertheless the state of the fire alternates between full-blast ("on") and quiet smoldering ("off"). If there were no heat exchanger, hot gases from the fire would carry too much heat away up the chimney during full-blast. On the other hand, if the heat exchanger cools wood or bituminous smoke too much, undesirable creosote is deposited in the chimney. There must be a compromise between efficiency and safety: The heat exchanger must be efficient, but not too efficient.

The Firebox. In the old days fireboxes were made of very heavy iron castings. But today all but a few are made of steel, which is cheaper and easier to work with.[2] Fireboxes intended for coal are supplied with shaker grates and lined with firebrick to delay burnout. Fireboxes intended for wood may or may not have grates and may or may not be lined with firebrick. The reason for these differences derives from the chemical and physical differences between the two fuels.

When wood is heated, water is given off and the wood cracks as it dries. Next, combustible gases evolve from these cracks. If hot enough, the gases burst into flame as they come in contact with oxygen above the solid wood. But coal, especially hard coal, does not contain much water and does not give off so much combustible gas. Rather, burning takes place at the glowing surface of the coal chunks. Chemically coal-burning is a union between carbon and oxygen at a gas/solid interface and not a union between oxygen and partially oxidized hydrocarbons in the gas phase. Thus, in burning coal, the problem is one of bringing oxygen into intimate contact with the surface of the chunks. Grates accomplish this by allowing air to flow up through the fuel bed. Because coal contains many mineral impurities, coal leaves much more ash than wood. The ashes impede the flow of air up through the gaps in the grates, especially if the ashes melt and fuse into clinkers. Thus coal grates must be shaken from time to time to restore air flow through the fuel bed. Triangular grates are common in coal furnaces, because they grind up clinkers as they are turned (**Figure 1.3**). The clinkers fall into the ash pit, leaving the unburned chunks of coal still on top of the grates.

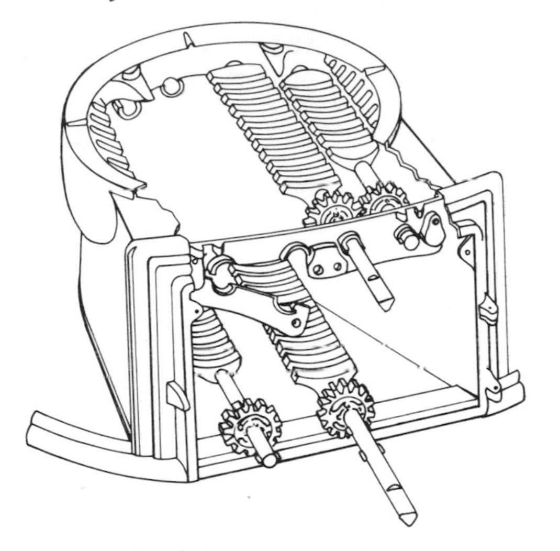

Figure 1.3 Coal grates can be shaken or turned. The triangular shape helps to break up clinkers.

Some wood furnaces also have grates, but the gaps are smaller, because wood ash is a fine powder that sifts into the ash pit with comparative ease. Another school of thought prefers burning wood without grates: (a) To avoid having to replace the grates when they burn out; and (b) To prolong the burn. Wood embers nestled in ashes stay alive and retain the capacity to ignite fresh wood for a long time—48 hours is not unusual. On the other hand, embers on grates tend to burn up or go out when the furnace is damped.

Grates in coal furnaces must be made of very heavy cast iron or they will soon burn out. The reason again can be traced to the nature of the burn. Surface burning liberates all the heat in a small space right at grate level where it may be contained because of the insulating effect of ashes. The flame of a wood fire, on the other hand, liberates heat over a larger volume. In technical terms, the power density tends to be quite a bit higher in a coal fire, and that is why coal grates must be rugged.

To make a point in the preceding paragraphs, we have treated the comparison of wood- and coal-burning as a contrast of black and white, when in fact there are many shades of gray between the extremes. Anthracite coal is almost pure carbon, and burning takes place essentially on the surface with almost no flame. But bitumi-

TABLE 1.3
Comparison of Solid Fuels

	Fuel value BTU/lb[a]	*% fixed carbon by weight*[b]	*% ash by weight*	*% sulphur*
Anthracite, Pennsylvania	13,000–14,000	80–85	9–12	0.7
Bituminous, moist, Ohio	12,000–13,000	40–50	8–12	1.5–3.0
Sub-bituminous, moist, Wyoming	9,000–11,000	38–48	2–6	1.0
Lignite, 30% moisture,[c] South Dakota	6,000–7,000	20–30	7–10	1.0
Wood, 20% moisture[c]	7,000	44	2	less than 0.1

[a] Heat of combustion per pound of fuel as fired, i.e., per pound of combustible material and minerals and water.
[b] Fixed carbon is that remaining after volatile matter has been driven from the coal by heating to 1750°F in a closed crucible.
[c] Moisture on a wet basis, i.e., 30 grams of water to 70 grams lignite.

nous and sub-bituminous coals burn with a long flame in the initial stages of the burn, just as wood does. As a consequence these soft coals can give rise to creosote and soot deposits in the chimney, just as wood can. Lignite is of course even closer to wood in chemical composition and fuel value (see **Table 1.3**), and therefore also in burning characteristics.

A generalization that follows from this discussion is that wood can be burned safely in any firebox intended for coal, although it may have to be cut into very short pieces. The converse is not true.

Attempts to burn the gases evolved from wood and soft coal completely have usually been along two lines: One has been to admit "secondary" air above the fuel to ignite unburned gases, but this always fails at low heat output rates because the gases must be fairly hot (around 1000°F) for ignition to take place at all. Recognizing this difficulty, various inventors as far back as Benjamin Franklin have designed the firebox so that unburned gases are forced down through a grate holding red hot coals to ensure a temperature high enough for ignition. Results have fallen short of the ideal, or we would all be heating with stoves and furnaces of downdraft design by now. **Figure 1.4** is a diagram of the Woodomat Furnace, devel-

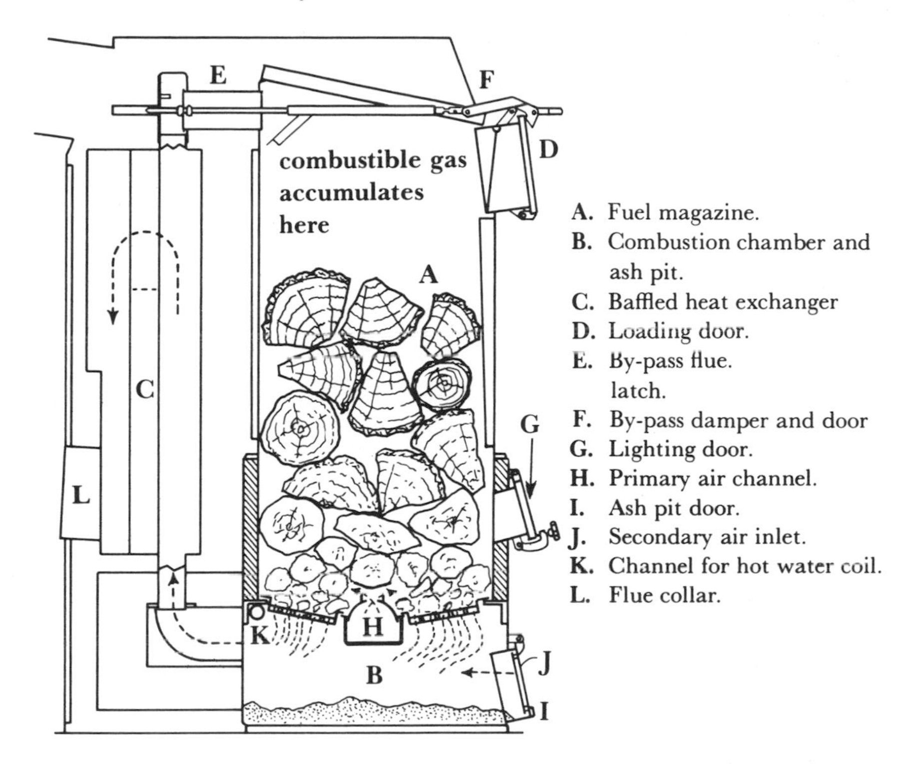

Figure 1.4 The Woodomat. Wood is loaded through the magazine door **D.** Primary air enters through passage **H** and secondary air through ash door **I. C** is a large heat exchanger. It worked, more or less after a fashion.

oped with the aid of the Northeastern Wood Utilization Council during the 1940s. Similar coal furnaces have appeared from time to time. One trouble with the Woodomat, along with other downdraft furnaces, is that unburned gases build up in the fuel magazine when the furnace is relatively cool. If the operator then opens the loading door abruptly, the unburned gases may mix with oxygen fast enough to cross an explosion limit. This surprises the operator. The trick is to open the door slowly from a distance. Franklin's advice on downdraft stoves was not to let servants handle them, as such stoves were inherently temperamental and required a degree of technical sophistication on the part of the operator.

Most solid-fuel furnaces today are of the tried-and-true updraft or cross-draft firebox design, not much different from the vast majority of their predecessors. They are steady, predictable workhorses. They tend to be smaller than furnaces of yesteryear because houses are smaller and better insulated today, and because many modern furnaces are intended as add-ons, that is, intended as adjuncts to already-installed fluid-fuel furnaces.

Controls. Flow of combustion air into the firebox is controlled by an upstairs thermostat, which opens and closes a shutter by means of a small motor or solenoid (**Figure 1.5**). When room temperature falls below the thermostat setting, the motor opens the shutter and air rushes into the firebox to stimulate the fire. Alternatively, combustion air may be provided by a small blower as in Figure 1.1. Generally speaking, blowers cause faster combustion and generate higher firebox temperatures. This ordinarily requires stouter firebox construction. One drawback of the combustion blower is that flow of air into the firebox is not cut off completely when the blower is off, because of residual flow through the fan housing. Thus control

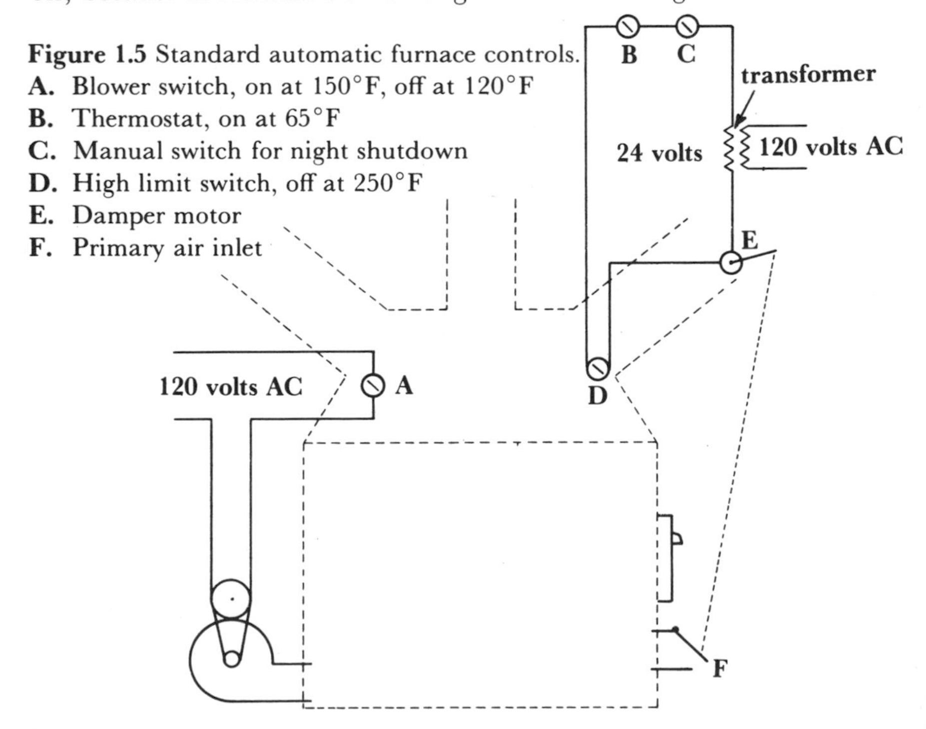

Figure 1.5 Standard automatic furnace controls.
A. Blower switch, on at 150°F, off at 120°F
B. Thermostat, on at 65°F
C. Manual switch for night shutdown
D. High limit switch, off at 250°F
E. Damper motor
F. Primary air inlet

over the fire is not as complete with a combustion blower as with a shutter. Manufacturers who have adopted the combustion blower believe it contributes to higher combustion efficiencies. This is plausible, but proof is lacking.

The distribution blower is controlled by a heat-sensitive switch in the bonnet that turns on the blower as the bonnet temperature rises above some setting, normally in the range 150°F to 175°F. The switch turns off the blower when the bonnet temperature falls below some lower limit, which is variable and can be set so that cycling is not too frequent. Switches incorporating different on and off settings are called *differential* switches. Optimum settings of these switches depend on the burning characteristics of the particular furnace in question.

All automatically controlled furnaces should have a high-limit switch that stops flow of combustion air whenever the bonnet temperature exceeds 250°F. This switch is normally in series with the thermostat. Its purpose is to prevent furnace and ductwork from overheating if the distribution blower fails for some reason.

Stokers. In hand-fired coal and wood furnaces, the flow of combustion air is controlled in accordance with heat demand. A stoker furnace is fundamentally different in that the thermostat controls flow of fuel into the firebox as well as flow of air. **Figure 1.6** shows a typical residential underfeed stoker which can be used with pea coal or wood pellets. When the thermostat calls for heat, a motor turns an auger that delivers coal or pellets to a heavy cast iron firepot. Simultaneously a blower goes on to force air into the combustion region through *tuyeres*, air passages in the firepot. As bits of fuel are pushed upward in the firepot, they get hotter and hotter until volatile gases are given off and ignited as they rise through the combustion zone. Then the remaining carbon burns as it subsequently passes through the hot zone, leaving ashes which either sift over the side of the firepot or fuse into clinkers which must be removed daily with tongs. During standby, stokers deliver a small amount of fuel to the furnace at regular intervals just to maintain a small pilot fire. Alternatively, fuel to hold the fire can be added to the firebox automatically whenever the firebox cools below some temperature limit.

It is entirely feasible to convert an oil or gas furnace to stoker

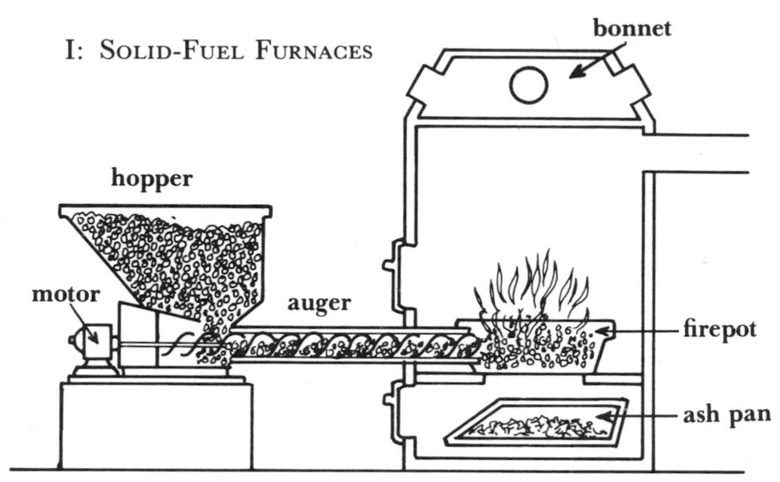

Figure 1.6 An underfeed stoker. On a call for heat the motor turns the auger and simultaneously blows air into the combustion chamber through *tuyeres* in the cast iron firepot.

operation as long as the firebox is big enough to accommodate the firepot. The furnace is more or less indifferent to what fuel the heat comes from, although cast iron boilers are somewhat problematical, since stresses on plates are often gauged according to temperatures to be expected from an oil or gas flame. One problem with such a conversion is that a stoker may not keep up with heat demand, since the oil or gas gun it replaces is usually a more powerful source of heat. Wood pellets and chips have an even slower heating rate than coal because of the difference in fuel value (see Table 1.3). This may be compensated for to some extent by adjusting the speed of the stoker and the rate at which air is blown into the firebox. Another problem with a converted firebox is that there is no room for an ash bucket under the firepot. Thus, ashes must be shoveled out—a dirty job—or an ash removal auger must be installed too. At regular intervals this pushes ashes out of the firebox and into waiting ash cans.

A potential stoker buyer should be aware that stokers designed for anthracite alone are not suitable for use with bituminous coal. Bituminous clinkers may form a hard crust over the firepot that prevents heat from escaping and may cause breakdown of the metal in the firepot. Bituminous stokers often have a rotating gear around the firepot that breaks up clinkers. See the end of this chapter for a list of stoker manufacturers.

Manual vs. Automatic Control. So far we have made it sound as though every solid-fuel furnace must be controlled automatically. The fact is that in well-insulated new houses a manually operated stove is adequate to provide all the heat desired. The stove can be in the basement and—with the addition of some simple ductwork making it into a furnace and perhaps qualifying you for a tax credit—heated air can be circulated very nicely by gravity. This way there is no noise, no dependence on electricity, and heat does not come in spurts; instead a gentle stream of warm air flows up continuously through one or more registers. The damper can be manipulated by you from upstairs with the aid of a chain. You become the thermostat.

By well-insulated we mean a house that is tightly sealed against drafts and that has the equivalent of six inches of fiberglass in the walls and perhaps a little more in the ceiling. A house that needs only 20,000 BTU/hr to sustain a comfortable indoor temperature, even when the temperature is below 0°F outside. Once the house is warm, only a little makeup heat is required to take care of the small rate of heat loss through the insulated shell. If the house has been designed well, it is a good deal heavier than standard, so that abrupt swings in temperature are eliminated. When the air begins to feel a little cool, you give a tug on a chain connected to the damper on the small furnace below.

There is something to ponder here. A manually operated furnace is usually damped just before bedtime. Overnight, creosote may be deposited in the chimney, and this may be ignited first thing in the morning when you get up and open the damper wide to warm the house in a hurry. The point is that you are in attendance when the likelihood of a dangerous chimney fire is greatest. With conventional automatic control, the thermostat can stimulate the fire any number of times during the night. On each firing there is some small probability of a chimney fire, but instead of standing by with an eye on the situation, you and your family are fast asleep. Thus it is plausible that manually operated systems are safer, unless the furnace is damped all night long by a manual switch in the thermostat circuit (Figure 1.5). If the furnace is left to automatic control overnight, extra diligence is required in keeping the chimney clean.

FEATURES TO LOOK FOR IN A SOLID-FUEL FURNACE

A person about to buy a furnace faces a bewildering choice today. There are at least 100 companies turning out many different models as fast as they can. Most of these companies weren't in existence when the oil embargo was imposed in 1973. Thus their experience is not exactly vast. No doubt some designs are seriously flawed. Others still haven't been completely debugged. Nevertheless, there is reason to be reassured. Building a furnace is not like building a space shuttle or a nuclear power plant. Intense competition has already weeded out some noncompetitive models and led to improvements in others. Excellent communication ensures that word of what works and what doesn't gets around the industry fast. Most manufacturers belong to either the Wood Energy Institute or the Solid-Fuel Heating Council, both of which keep their members informed of new developments through trade publications and meetings. The furnace industry has come on strong in a very short time.

The above words offer reassurance that chances are good that you will end up with a decent piece of hardware in the basement. Nevertheless, here are some pointers to help guide your purchase.

Firebox. To some people there is nothing more irksome than having to cut wood up into small pieces. If you're planning to burn wood, it's a good idea to get a furnace that takes a two-foot log that's on the long side. This implies a 30-inch firebox. A two-foot log is about as long as is convenient to handle. The volume of the firebox should depend on how much heat you need and what fuel you are planning to burn. The calorific value of hard coal is about 650,000 BTU per cubic foot, whereas for white pine it's only 186,000 BTU per cubic foot. If you have only pine to burn, you will need a large firebox unless you plan to spend most of your time in the basement.

There are two schools of thought on whether the firebox of a wood furnace should be lined with firebrick. If combustion air is provided by a blower, then firebrick is needed because of the high temperatures achieved in the firebox. This is hard combustion. On the other end of the spectrum is soft combustion characteristic of most wood stoves, many of which do not have and do not need firebrick or other liners. With the price of steel rising rapidly (going up in 1979 at a rate of over 25 percent per year), the trend will be toward fireboxes made of thinner steel and lined with brick, cast iron or even air-cooled steel liners. Some furnaces are now available with stainless steel liners. One advantage of building with fairly light steel and using replaceable liners for protection is that the knocked-down furnace is easier to move into the house. The trouble with firebrick is that it needs replacing from time to time. If you're inclined to heave logs in, then look for a furnace with steel plate protecting the firebrick in the back of the firebox.

Most fireboxes intended for coal are lined with firebrick. An unlined firebox intended for wood may not be adequate to withstand the higher temperatures produced by coal. In a coal furnace, look for heavy grates and a wide loading door that you can get a shovel through. A good feature of some doors is that the handle operates on cam action, so that a good seal is maintained even when the latching mechanism becomes worn. Some manufacturers provide adjustable hinges, which allow you to reposition the door if it becomes loose through wear and tear. A tight seal around the door is desirable. This can be achieved with a gasket or by careful fitting of metal to metal.

As a historical note, many cast iron furnaces in the old days were not air tight. To damp these furnaces, a "check draft" on the flue was opened at the same time the primary damper was closed. This way air was drawn into the hot chimney directly from the basement without going through the firebox and the fire was held under control. The tradeoff was loss of warm basement air and infiltration of cold outside air to replace it.

Insulation. The furnace jacket should be insulated, unless you plan to heat the basement. On most modern furnaces there is about one inch of fiber glass, usually with a shiny layer of aluminum facing inward, on the inside of the casing. Two inches can be justified economically.

Heat Exchanger. It will be a great nuisance to you if the heat exchanger cannot easily be inspected and cleaned. Depending on where the furnace is set up in the basement, the cleaning ports are usually most accessible if in front.

And now a word of caution. There are some furnaces on the market whose heat exchanger is located in that part of the furnace normally connected to the cold-air return. This is not a good idea when the cold-air ducts in the house are built partially of wood, as is often the case. The problem is that the air circulation may reverse and carry heat from the heat exchanger up into the cold-air return when the power is off. This problem may not be a serious one, in fact it may not be responsible for even one house fire a year, but it is surely not good practice to build the heat exchanger into the cold-air return. The arrangement in Figure 1.1 is far preferable. There, heated air rises in steel hot-air ducts by gravity if electric power fails.

Controls. The wiring should be out of harm's way where it won't get hit and where it won't get too hot.

Filters. Can the furnace be outfitted with filters? Wood and coal are dustier fuels than you are accustomed to.

Domestic Hot Water. The cost of heating domestic water with gas or electricity is going up fast. Why not get a coil in your furnace to give you relatively inexpensive hot water during the winter? The coil and associated plumbing will pay for itself in two or three years. Chapter 5 treats this subject in detail.

Humidifier. Well-sealed houses do not need to be humidified in winter, but drafty ones may. The solution to the problem of dry air is to caulk and weatherstrip to keep dry winter air from coming in and to give natural sources of moisture in the house a chance to humidify the air before it leaks out. Humidifying by evaporating water costs energy; humidifying by weatherstripping and caulking saves energy.[3] Likewise, storm windows help to maintain high internal humidity by raising the temperature of the inner pane and thwarting Jack Frost.

To quantify this a bit, the Handbook of the American Society of Heating, Ventilating and Air-Conditioning Engineers (ASHRAE) now sets 20 percent relative humidity as the minimum for comfort regardless of temperature. (This is a cultural, time-dependent matter. In 1948 the ASHRAE *Guide* recognized 30 percent relative humidity as the minimum value consistent with comfort. Many people are convinced that extremely dry saunas have a pronounced beneficial effect on health.) For the purposes of this example, assume that you want to keep the house at a temperature of 70°F and a relative humidity of 20 percent. Then the air turnover rate in a house of 10,000 cubic feet must be no more than one-third complete air change per hour, assuming an outside temperature of 10°F and outside relative humidity of 100 percent, and assuming that ordinary household sources of water vapor—cooking, bathing, house plants—humidify at a rate of one-half pound water per hour. Drafty houses can have exchange rates of more than two per hour. On the other extreme, the exchange rate can be as low as one-twentieth per hour in a super tight house, in which case dehumidification may be necessary. A reasonably well-built house of 1950 or 1960 vintage will probably have an air turnover rate of about one-half complete change per hour, depending on outside temperature, wind velocity, and conscientiousness of the building contractor.

Warranty. Take a look at the furnace's warranty. It probably more or less reflects the manufacturer's confidence in his product. Just because there is a warranty, however, doesn't guarantee that you can collect if something goes wrong. There are ways of stonewalling the plaintiff. Probably more important than a legally phrased piece of paper is the reputation of the dealer or heating contractor from whom you buy the furnace. If he is highly regarded in the community, it's no doubt because he earned the reputation by standing behind the products he sells.

Efficiency. There is no way that you can know whether furnace A is more efficient than furnace B, that is, which produces the greater heating effect per unit wood or coal. Nor do you necessarily *want* the more efficient furnace. This may sound unreasonable, so here is the argument.

If you have an old chimney which is unlined and/or not in good condition, it may be the better choice to sacrifice some heat just to keep the chimney clean. Bituminous coal, peat, lignite and wood all give rise to gases that may condense into creosote in the chimney. No furnace currently available (with one exception, see page 53) burns the gases so completely under all conditions that there is no potential for creosote formation. One way to prevent creosote from forming is to keep the temperature of the flue gases above the condensation point of creosote, which is in the range of 100–300°F, depending on the relative amounts of oxygen, nitrogen, water and all the organic components of smoke. However, this heat that keeps the chimney free of creosote is not available for inside heating. Efficiency has been sacrificed. Thus, one may decide to buy safety at the expense of efficiency.

The safety issue aside, you still do not know the measured efficiencies of the furnaces you may be comparing. Neither does anyone else, at least not precisely. They have not been tested under laboratory conditions. What can be said with some confidence is that the thermal efficiency of wood furnaces is probably about the same as wood stoves, which have been shown to be 50–65 percent efficient.[4] Coal furnace efficiencies have been measured in the same range.[5] There is very little doubt that most manufacturers have—through intimate experience with their products—made them nearly as efficient as can be without a radical change in the technology. One should regard with suspicion claims of technological breakthroughs that result in dramatic superiority in heating effect. The assertions are written for the most part by admen without sound technical data to back up their extravagant statements.

To some extent you control the efficiency of your furnace. In all forced-hot-air systems there is a distribution blower. By varying the speed of the blower through use of pulleys, you can alter the efficiency with which heat is extracted from the heat exchanger. A reasonable approach is to set the speed of the blower so that the gases leaving the furnace are only so warm that condensation does not occur in the chimney; though, even so, there will be some condensation during the smoldering phase of the burning cycle.

SIZING THE FURNACE

It can be a mistake to decide on the size of furnace for your house by using the manufacturer's power rating in BTU/hr as a guide, because such heat output ratings assigned to solid-fuel furnaces by manufacturers are crude estimates at best, and not even based on a standard set of assumptions. Some manufacturers' ratings are estimated average heat outputs for some hypothetical house under some arbitrary set of weather conditions. Others' ratings are estimates of the maximum the furnace will deliver with the best fuel and the highest draft. Furnace ratings reported in terms of cubic feet or square feet of heated area are even more meaningless than those reported in BTU/hr, because they make assumptions about how well the house is insulated. No one is trying to hoodwink you. It's just a fact of life that the born-again American solid-fuel furnace industry is not yet as technically sophisticated as it will be given a few more years. Fortunately, heat output ratings are not terribly important as long as you can find a knowledgeable dealer to talk to. The best datum you can give him is the number of gallons of oil or number of cubic feet of natural gas or kilowatt-hours of electricity it took to heat your house last winter. In heating effect, one cord of good hardwood or one ton of coal is equivalent to about 160 gallons of fuel oil. Thus the dealer can estimate the number of cords of wood or tons of coal you will burn, and this will tell him pretty well what size furnace you will need. (For rough estimates, again, one cord of the best hardwood is equivalent to two cords softwood, one ton coal, 160 gallons oil, 260 gallons of liquified propane, 1400 cubic feet of natural gas, 7000 kilowatt-hours.)

The choice of size for your new furnace is not crucial where the previous furnace is left for backup, since the old equipment can take over if necessary. Many dealers advocate—sometimes as if proclaiming received wisdom—undersizing the solid-fuel furnace so that long hold-fire periods are eliminated and creosote production minimized. There are two drawbacks to this approach. One is that a more costly source of heat must be called upon to augment the solid-fuel furnace in cold weather. If electricity is used for backup, it will be needed just at the wrong time, namely on the coldest days when the strain on the electric power grid is already at its peak. The other trouble with undersizing the solid-fuel furnace is that refueling times will be short. Those who go away from nine till five will want a furnace large enough to carry the house unattended for at least eight hours in the coldest weather.

In more than one case the size of the furnace has been determined by the width of the basement door, although one dedicated wood-heat enthusiast, rather than allowing himself to be intimidated, knocked out part of his foundation just to get the furnace he wanted where he wanted it. See Appendix 1 for a furnace-sizing method based on previous oil or gas consumption.

COMBINATION FURNACES

Dual-fuel or "combi" furnaces are of several types. The simplest can accommodate either coal or wood. The firebox of such a furnace is usually lined with firebrick to protect the steel against the high temperatures from burning coal. Conversion from coal to wood is made by laying a perforated plate over the coal grates to support the finer ash from wood.

Another type of combination furnace is composed of two separate fireboxes, one for oil or gas and the other for wood and/or coal. Usually there is a single heat exchanger as shown in **Figure 1.7.** There are two advantages to such combination units when compared to two entirely separate furnaces: (1) They take up less space; and (2) Connections to ductwork are simplified by unified design. However, both fireboxes in the combi tend to be on the small side. Corollary: you may have to get up at night to refuel. Where there is already a decent oil or gas furnace and space is not limiting, the sensible course is to add a coal or wood furnace and not to replace

Figure 1.7 The Hunter wood-oil furnace with separate combustion chambers and a single heat exchanger.

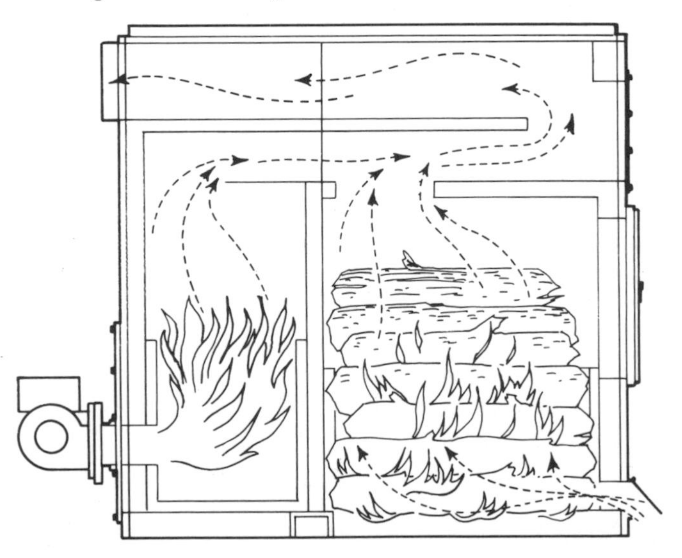

the original with a combi. On the other hand, in new well-insulated houses with low heat demand a combi may be an economical and very satisfactory choice.

There is a third type of dual-fuel furnace intended mainly for solid fuel, but with a gas or oil burner mounted in the same firebox. This arrangement has the drawback that the oil or gas gun may become clogged with creosote from the wood or coal fire. Generally speaking, this kind of combi is less satisfactory than the type with separate fireboxes.

SAWDUST AND CHIP FURNACES

There are several makes of sawdust furnace available today.[6] They tend to be on the large side, as they are mainly intended for heating mills where sawdust is a byproduct of production. The Rettew, shown in **Figure 1.8**, is made in three sizes ranging from 500,000 to 1.5 million BTU/hr. The nice thing about sawdust is that it, unlike cordwood, can be handled automatically. A truck drives into your driveway once a month and blows sawdust into a bin in your base-

Figure 1.8 The Rettew sawdust and chip furnace. The vibrator keeps fuel flowing into the combustion chamber.

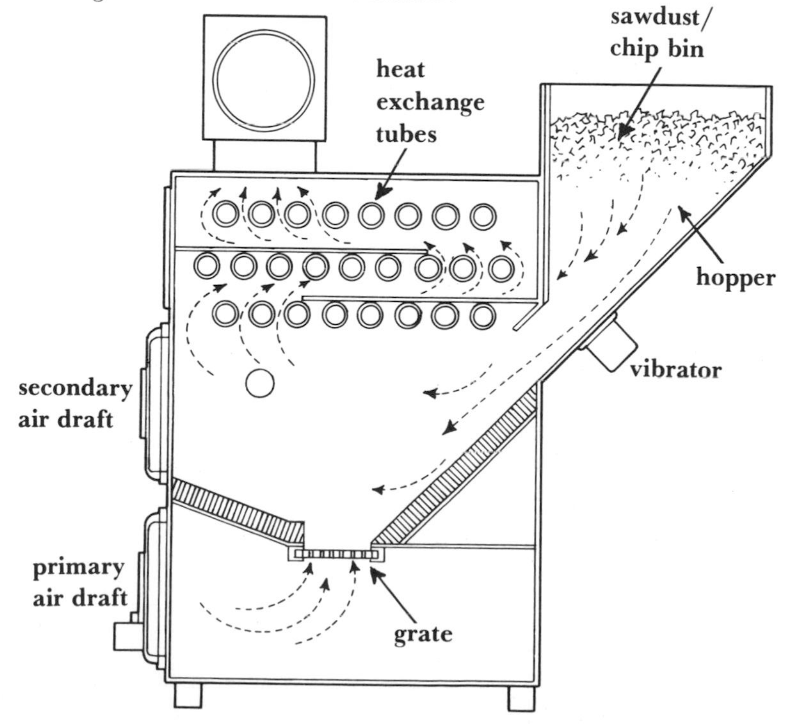

ment. The bad thing about sawdust is that it packs together and requires periodic fluffing or shaking for good combustion. Fuel in the hopper may be replenished by hand or by an automatic auger.

The Conifer, another chip furnace which still can be found in many residential basements on the West Coast, is now being made by American Fyr-Feeder Engineers. Large fireboxes and hoppers may be purchased separately from Fyr-Feeder for retrofit onto preexisting furnaces and boilers. Sizes range up to 8 million BTU/hr, enough to heat a factory.

INDUSTRIAL FURNACES

Many industries produce a good deal of scrap wood as a byproduct of the manufacturing process. There are furnaces available big enough to heat these factories using only scrap wood. An example is shown in **Figure 1.9.** This is a furnace rated at 1.3 million BTU/hr by its maker, G and S Mill of Northborough, Massachusetts. Such a large furnace can be fed by forklifting scrap into the firebox through a 34-by-24-inch door. One load equals one-half cord.

Figure 1.9 A G & S Mill furnace with firebox and heat exchanger separate. The heat exchanger is so effective that water in the smoke is condensed and drained out at the bottom.

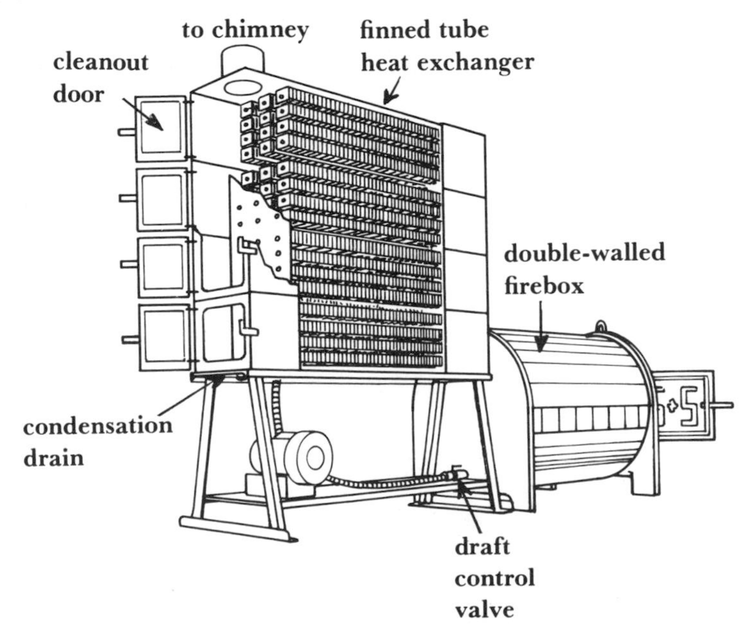

G and S Mill furnaces exhibit considerable originality. The firebox of the 1.3 million BTU/hr furnace is a steel cylinder with a wall 3/4-inch thick. Surrounding the first cylinder is a second cylinder of stainless steel with an air gap between where combustion air is prewarmed before it is blown into the firebox. Thus heat lost through the firebox walls is reclaimed and the firebox temperature is kept high for good combustion. Paul Kalenian, the inventor, is interested in efficiency, so he has coupled the firebox to a finned-tube heat exchanger that is so effective that steam in the exhaust gases is recondensed and the latent heat of vaporization of water recaptured. The tradeoff is frequent brushing of the tubes in the heat exchanger to remove creosote deposits. The furnace in Figure 1.9 weighs 13,800 pounds; both smaller and larger versions are available.

CODES

Some states, Massachusetts for example, require that furnaces be listed by a recognized testing laboratory. Testing and listing of furnaces is not as extensive as it is for stoves yet, but it soon will be. Massachusetts recognizes approval of furnaces by ETL, the Energy Testing Laboratory of Maine. Underwriters' Laboratories are now in the process of developing a testing program for furnaces.

How important is approval by such a laboratory? This is hard to answer because there is not enough experience to go on with solid-fuel furnaces. The best one can do is extrapolate from what is known about wood stoves, for which there is a data base. In the vast majority of cases house fires connected with wood stoves are the result of faulty installation, faulty chimney, or carelessness on the part of the operator, and not the result of a defect in the stove.[7] Thus, a fair conclusion is that having an approved stove or furnace is not very crucial to the homeowner as far as safety goes, but it may be of some importance in relations with building inspectors and insurance agents. Testing drives the price of stoves up, and drives small manufacturers out of business without having much impact on the number of house fires.

There are instances where stove testing has had an effect exactly opposite to the one hoped for. Part of the testing requirements specify temperature limits in the flue when the stove is run full tilt. To meet these limits manufacturers reduce the size of air inlets, which reduces the responsiveness of the stove and its maximum heating rate. One way people compensate for this degradation of performance is to run the stoves with the loading door open. This is a good way to start a chimney fire.

All of this is not to say that listing of stoves and furnaces is always counterproductive or just an empty exercise. It is possible that the overall effect of testing on safety may be positive. But it is not obviously so, and all the evidence indicates that emphasis should be on making sure the chimney is sound and clean and on a scrupulous adherence to safety rules in installation. Finally, keep your wits about you and don't do stupid things like drying clothes on the furnace bonnet.

LIST OF STOKER MANUFACTURERS

American Coal Burning and Woodstoker Corporation, 1133 W. Cornelia Avenue, Chicago, Illinois 60657

Applied Engineering, Orangeburg, South Carolina 29115

Axeman–Anderson, 233 West Street, Williamsport, Pennsylvania 17701

Babcock and Wilcox, 1 Northfield Plaza, Troy, Michigan 48084

Canton Stoker Corporation, 307 Andrew Place S.W., Canton, Ohio 44706

Carp Heating and Air Conditioning, 2135 St. Clair Avenue, Cleveland, Ohio 44114

Chicago Stoker Corporation, P.O. Box 68, 704 North Wolf Road, Wheeling, Illinois 60090

Dayton Automatic Stoker, P.O. Box 255, No. Dayton Station, Dayton, Ohio 45401

Detroit Stoker Company, 1510 East 1st Street, Munro, Michigan 48161

General Machine Corporation, Emmaus, Pennsylvania 18049

Illinois Stoker Company, 102 W. 7th Street, Alton, Illinois 62004

John Rogers Company, P.O. Box 20232, Louisville, Kentucky 40220

Kramer Brothers Foundry Company, 17 Dell Street, Dayton, Ohio 45404

Laclede Stoker Company, Tower Grove Foundry, 4440 Hunt Avenue, St. Louis, Missouri 63110

Leffel and Company, Springfield, Ohio 45505

Riley Stoker Corporation, 100 Neponset Street, Worcester, Massachusetts 01606

S and S Manufacturing, 3956 Highway 119, Longmont, Colorado 80501

Therm–Kon, Galesville, Wisconsin 54630

Van Wert Stoker Manufacturing Company, Peckville, Pennsylvania 18452

Will–Burt Company, Orrville, Ohio 44667

Worley Equipment, Inc., 2301 N. Knox Avenue, Chicago, Illinois 60639

NOTES TO CHAPTER I

1. In New England the price of oil has gone up at about the general inflation rate over the last decade, except for big jumps in 1974 and 1980. The price of wood in New England seems to be increasing at about the general inflation rate. Coal, which is intermediate in price between oil and wood in most areas, will probably continue to cost less than oil, but may go up in price at about the same rate as oil. Wood is not tied into the national economy to nearly the same extent as coal. Thus wood prices will be determined in large degree by local supply and demand.
2. There is a popular misconception that cast iron is used in stoves and furnaces in preference to steel because it "holds its heat". The fact is that the heat-storage capacity of both is 0.12 cal/gm-°C. Gram for gram or pound for pound there is no difference.
3. J. W. Shelton, "The Energy Cost of Humidification", *ASHRAE Journal,* January 1976, page 52.
4. J. W. Shelton, T. Black, M. Chaffee, M. Schwartz, "Woodstove Testing Methods and Some Preliminary Experimental Results", *ASHRAE Transactions,* 84 (1978), pages 388–404.
5. J. R. Allen, "Heat Analysis of a Hot-Air Furnace", *ASHVE Transactions,* 22 (1916), page 293. A. P. Kratz, "Performance of a Warm-Air Furnace with Anthracite and Bituminous Coal". *ASHVE Transactions,* 30 (1924), page 277.

6. American Fyr-Feeder Engineers
1265 Rand Road
Des Plaines, Illinois 60016

Gaulrapp Sheet Metal
4 Widgery Wharf
Portland, Maine 04111

Rettew Automation, Inc.
Box 65, North Sheridan Road
Newmanstown, Pennsylvania 17073

7. J. W. Shelton, "Analysis of Fire Reports on File in the Massachusetts State Fire Marshall's Office Relating to Wood and Coal Heating Equipment". Publication NBS-GCR-78-149 (1978), sponsored by the National Bureau of Standards and the U.S. Department of Energy.

CHAPTER II

Furnace Installation

Selecting a furnace solves only part of your heating problem. An equally important part is how to integrate the furnace into your house for safe and satisfactory performance over the years. The chimney is absolutely critical to the successful operation of a coal or wood furnace.

CONNECTING THE CHIMNEY AND DUCTWORK

In the author's experience, insulated stainless steel chimneys outperform conventional masonry chimneys when the fuel is wood or coal.[1] With insulated stainless steel chimneys, the furnace is more responsive, top heat output is apt to be higher; soot and creosote deposits are apt to be much less troublesome. By burning the furnace very hot for fifteen minutes once a day it is often possible to eliminate altogether the need for mechanical chimney sweeping of a stainless chimney. As with any chimney, careful surveillance of the inside of the metal chimney is an absolute rule, since a very hot, prolonged fire in a stainless steel chimney can cause it to pop apart at the seams and buckle. The easiest way to inspect a stainless chimney is with a mirror inserted through a tee at the base of the chimney. If creosote buildup is greater than one-eighth inch, it's time to clean the chimney.

The Chimney Connection. The cross-sectional area of the chimney flue should be equal to or greater than the area of the flue collar of the furnace (see Figure 1.1). The furnace manufacturer has found through experience the flue size that works best with his product, and the collar size may be taken as his recommendation of minimum flue size. A nearly absolute rule is that reduction of flue size below the manufacturer's recommendation will lead to poor performance—that is, to smokiness and a reduction of heat output.

It usually does no harm to *oversize* the flue by an inch or two, but creosote deposition and loss of draft can result if the flue is oversized by much. In masonry chimneys exhaust gases near the wall are considerably cooler than those at the center. In effect there is a chimney within a chimney, hot gases rising in the middle and cooler, denser gases falling near the wall. This counterflow sets up swirling eddy currents that spoil the draft.[2] The larger the cross-sectional area of the chimney, the worse this effect is apt to be.

Avoidance of Creosote. The reason why creosote buildup is generally slower in insulated stainless steel chimneys than in masonry chimneys is because steel chimneys keep the stack gases hotter. The one inch of mineral wool in these chimneys has an R-value (thermal resistance) of about 3 ft^2-°F-hr/BTU, whereas a typical tile-lined 4-inch brick chimney wall has an R-value of about 1.5 at low stack temperatures, assuming an air gap of ¼ inch between the liner and the brickwork. The R-value of the brick chimney decreases at higher stack temperatures, primarily because the air gap poses less and less of a barrier to heat flow, both by convection and radiation, as the temperature rises. During a chimney fire the tile may get red hot, in which case heat jumps the air gap almost as if it weren't there.

The cylindrical geometry and shiny surface of an insulated stainless steel chimney also favor lower heat loss. The interface between the outer surface of the chimney and surrounding air is a barrier to heat flow. The larger the area of this interface, the greater the heat loss from the chimney. A stainless steel chimney with a 7-inch inside diameter has an outer surface area of 2.35 ft^2 per foot of length. The most common alternative with a nominal 7-by-7-inch flue liner and a single course of brick has an outer surface area of 5.33 ft^2 per linear foot, more than twice as great. Heat loss from the chimney is also dependent on the nature of the chimney-air interface. The shiny, smooth surface of the stainless chimney is a more effective barrier to heat flow than the rough surface of a brick chimney.

Both kinds of chimney perform better if internal rather than external to the building shell—stack gases stay hotter in them and heat lost from them helps to warm the house. There is one major advantage to a masonry chimney: it provides thermal ballast that tends to even out fluctuations in house temperature. This effect is not very important for a single-course brick chimney, but it may be if the chimney is truly massive. A minor advantage of the brick chimney is that it can provide enough heat in upstairs bedrooms to make them comfortable for sleeping.

Thus, other things being equal, the insulated stainless chimney does a better job—there is less creosote and draft is better. As we have observed, many people find that by firing the furnace at maximum output once a day, creosote buildup can be prevented in a stainless chimney. Any creosote deposits either flake and fall or are actually burned in what amounts to small, controlled, preventive chimney fires that stave off larger, more dangerous fires later. It is not unlike forestry practice in stands of southern pine where naturally ignited fires are allowed to burn up forest debris every few years. Former practice was to extinguish such fires with the result that forest litter accumulated and fires that did get out of hand were truly devastating.

Cleaning brick chimneys by the once-a-day hot-fire method is much less effective than cleaning stainless chimneys by this method. It must again be emphasized that the only way to know whether creosote buildup is excessive is to inspect the chimney regularly. Once a week is not too often until you know how your particular furnace-chimney combination behaves.

There is some evidence that certain salts—the chlorides of zinc, tin and copper for example—can reduce chimney deposits from bituminous coal when thrown on the fire from time to time.[3] Unfortunately, a systematic study of the effect of such salts on wood creosote has not been done. Mechanical chimney sweeping is tried and true. It can be very hazardous business, however, so leave it to a professional if the roof is high and steep.

Anticipating Chapter 4 a bit, there is a completely different strategy to avoidance of creosote, to wit: uncoupling heat production from heat demand by incorporation of heat storage capacity into the system. This allows long, hot burns during which heat is transferred to storage and eliminates the smoldering phase of the burn altogether. During the burn the stack temperature should be kept high enough to preclude condensation of creosote in the chimney. Heat is then blown into the house when the thermostat calls for it. In **Figure 2.1** a rock storage bed receives heat either from a solar collector or a furnace. The one-way flaps in the two loops prevent reverse gravity flow of heated air from storage to the collector or furnace.

The question most frequently asked about installing solid-fuel furnaces is: "Can I safely use the flue my oil or gas burner is connected to for the solid-fuel furnace too?" The answer to this can be quite long-winded. A shortened version is: Many people do it successfully, but most safety officials frown on the practice. Some states have taken a stand against dual use of a single flue, while allowing combination furnaces, most of which are vented through a single flue. This kind of confusion is no doubt unavoidable when an unfamiliar technology is introduced as rapidly as solid-fuel heating today.

Two furnaces connected to the same flue can interfere with one another in a number of ways. For instance, addition of a second furnace might reduce the draft in the oil or gas burner and cause a loss in efficiency. A more serious interaction would be lack of ade-

Figure 2.1 Creosote formation may be avoided altogether when heat storage is incorporated into the system.

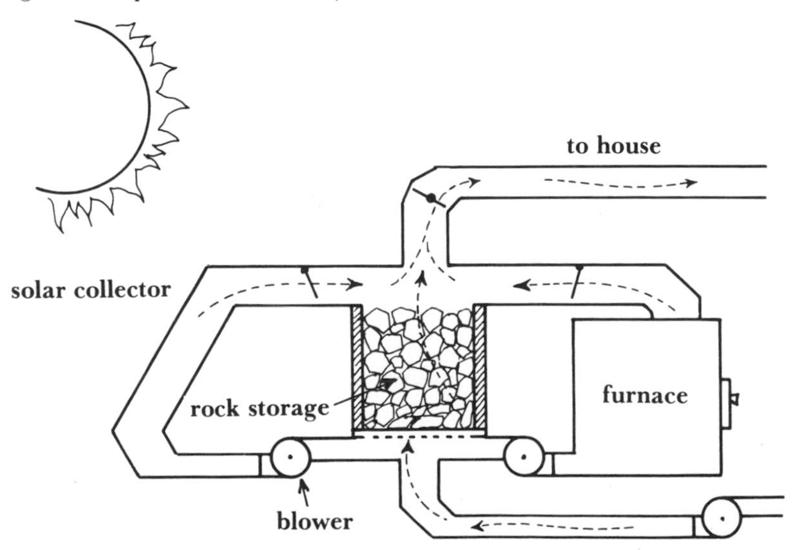

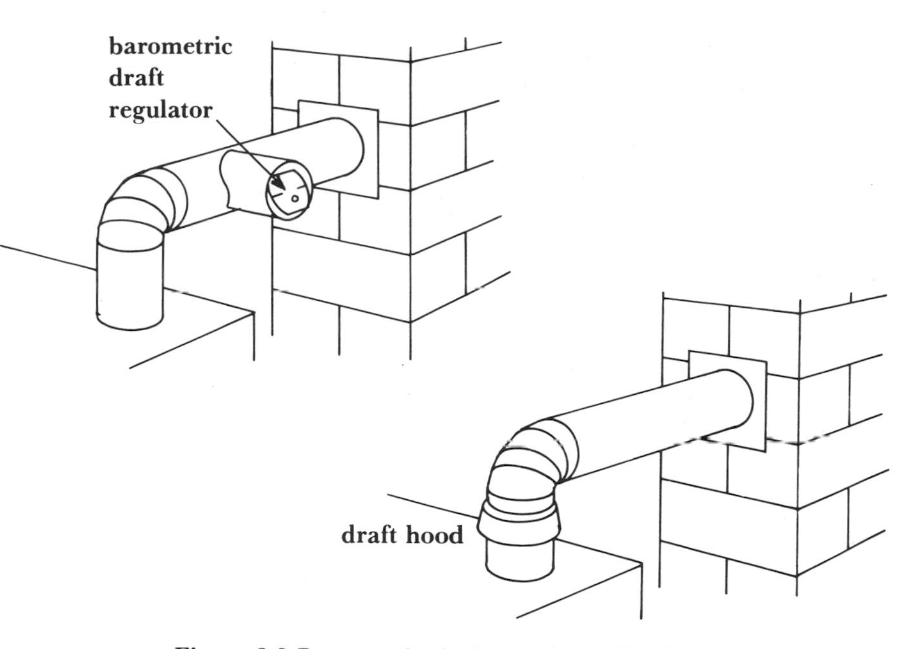

Figure 2.2 Barometric draft regulator. Draft hood.

quate chimney capacity to carry away all the products of combustion if both furnaces were going full blast at the same time. A chimney might have adequate capacity for two furnaces when clean, but not after creosote had built up in it and constricted the flue.

Another dangerous interaction is introduced by the barometric draft regulator used with the oil furnace or the draft hood of the gas furnace (**Figure 2.2**). The barometric draft regulator stabilizes the draft in oil furnaces at the design value and the purpose of the draft hood is to prevent a downdraft from extinguishing the pilot light in a gas furnace. The trouble is that during a chimney fire they allow an unlimited flow of air to enter the chimney and stimulate the fire. Thus, in case of a chimney fire in a flue to which an oil or gas burner is attached, the first thing to do is to cover the barometric damper or draft hood with aluminum foil to cut off air to the fire. The next thing to do is to call the fire department.

As usual, there are two sides to every story. One fire marshall, who was at first skeptical, but inclined to experiment instead of speculate, hooked up a small wood furnace to the flue serving his oil burner. What he unexpectedly found was that the creosote problem completely disappeared and that the two furnaces functioned well together in a kind of synergy. He attributed the lack of creosote to the fact that the oil burner came on only when the wood fire was smoldering, i.e., at precisely the time when creosote deposition would otherwise be most likely. The oil burner kept the chimney clean by maintaining the temperature of the flue above the condensation point of water and other constituents of smoke. Others have had similar results. It is of course wrong to generalize from limited experience. However, one possible conclusion, which should be thoroughly tested, is that statistically there may be a correlation between dual use of a flue and reduction in frequency of chimney fires.

In localities where dual use of single flues is permitted, it is usually recommended that the smokepipe from the solid-fuel appliance be connected to the flue below the smokepipe from the gas or oil furnace, for two reasons: (1) to maintain adequate clearance of the solid-fuel smokepipe from wooden joists above (see below); and (2) to prevent creosote from building up in the ash pit and blocking flow of combustion products from the oil or gas furnace (which would lead to inefficient combustion and possible spillage of combustion products from the fluid-fuel furnace into the basement). Safety is greatly enhanced in such installations *if* the flue is big enough to handle both appliances simultaneously, and the flue is kept clean.

ADVICE: If adding a wood or coal furnace to a pre-existing fluid-fuel furnace, check with your local fire marshal and insurance agent about dual use of a single flue. Their views on the subject are apt to be of most importance to you personally.

Smokepipe. Connecting a furnace to a chimney is in principle no different from connecting a stove to a chimney. The same safety regulations must be followed and the crimped end of the pipe should be positioned down so that goo cannot seep out. Sections of smokepipe should be fastened together with sheet metal screws, and the joints should be tight. The lowest section of smokepipe should be fastened to the collar of the furnace with sheet metal screws. This is to keep the smokepipe from coming apart in case a small explosion occurs in the furnace. These events are rare, but do happen occasionally in solid-fuel furnaces when hot combustible gases in the firebox are abruptly mixed with sufficient oxygen. If you buy a thin smokepipe you will learn in a year or two that heavier pipe is well worth the extra money. A local sheet metal shop may be able to make a smokepipe for you from stainless steel for extra long life.

One practical difference between stove and furnace installations is that the furnace chimney connector should almost invariably be as short as possible, whereas with stoves a moderately long connector is often desirable to improve the thermal effectiveness of the system. Most furnaces are built with efficient heat exchangers, and it is therefore unnecessary to use the smokepipe as a heat exchanger. Furthermore, furnace chimneys normally extend higher above the firebox, and—other things being equal—this aggravates the creosote problem and argues against a long smokepipe.

It is a good idea to keep an eye on the temperature of the stack gases by means of a thermometer with a three- or four-inch probe extending into the middle of the smokepipe. (Thermometers without a probe that strap onto the pipe are apt to give low readings.) If the temperature is routinely above 400°F, heat is probably being lost up the chimney unnecessarily. The idea is to keep the temperature of the gases as low as possible for efficiency's sake without condensation of creosote during the on phase of the burning cycle. To decrease stack temperature you can increase the speed of the distribution blower, or you can partially block the smokepipe with a manual damper.

ADDING A FURNACE

A solid-fuel furnace may be added to an oil or gas furnace so that they both use the same ductwork and in such a way that the original furnace takes over automatically if the solid-fuel furnace goes out. There is nothing intrinsically different about add-on furnaces, except they tend to be on the small side. Any solid-fuel furnace may be installed either in parallel or in series with another furnace (**Figure 2.3**). In the series arrangement only one distribution blower is required, and this may be the one already attached to the oil or gas furnace. Which arrangement to choose depends to a large extent on which one requires the least amount of work rearranging the ducts. Each installation is different, and therefore neither arrangement can be given general preference. Both work.

There is one very important point to observe in the series hookup,

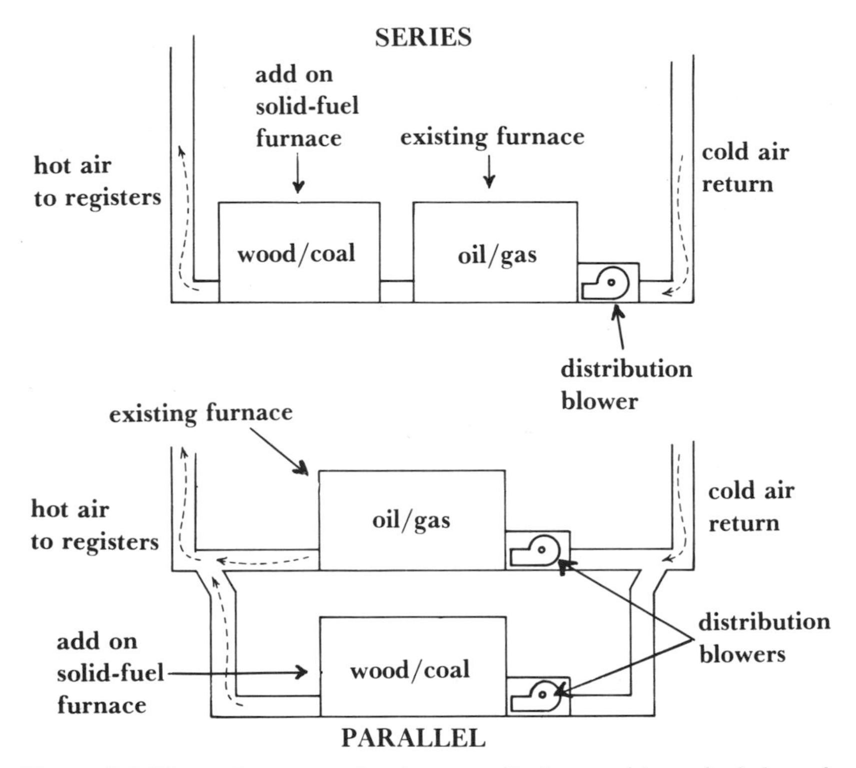

Figure 2.3 The series connection is generally less problematical than the parallel hookup. Note that the blower must be upstream of both furnaces in the series arrangement.

and that is that the positions of the two furnaces must be as shown in Figure 2.3 and not interchanged.[4] If they were, the blower would be exposed to the heat from the solid-fuel furnace and might burn out. Also, a wood or coal firebox upstream of the blower would be subject to reduced pressure that could draw smoke out of a leaky firebox and into the house. In the series hookup there are two thermostats upstairs (sometimes inside the same box) with the solid-fuel thermostat set higher than the oil or gas thermostat. The solid-fuel thermostat calls for heat by opening the draft regulator or turning on the combustion blower. The distribution blower is controlled by two independent heat-sensitive switches, one in the bonnet of either furnace. When the temperature in the bonnet of either furnace reaches 150–180°F, hot air is blown into the house.

In the parallel arrangement the controls are the same, except that each bonnet switch controls its own blower. If the parallel solid-fuel furnace is very small, it is sometimes possible to omit the duct between the cold-air return and the solid-fuel furnace (semi-parallel arrangement), since air from upstairs may come into the basement down stairwells and through cracks in the floor fast enough to keep pace with the blower of a small furnace. However, in the semi-parallel arrangement warm air from upstairs mixes with basement air, which represents a waste of heat if you don't need a warm basement. Another difficulty with the semi-parallel arrangement is that it may cause smoke reversal in the fluid-fuel furnace. Note the angles at which the new and old ducts are joined in Figure 2.3. By doing it this way you avoid the possibility of air heated by one furnace short-circuiting in the wrong direction through the other.

One nice thing about having an oil or gas furnace in the basement for backup, whether in series or in parallel, is that you can rely on it in fall and spring when the weather is relatively mild. You might find that a good deal of creosote would be deposited during mild weather if solid fuel were used because the furnace would be smoldering most of the time. In cold weather the chimney may stay warm enough that creosote buildup is not excessive.

A word of caution is appropriate here. The solid-fuel furnace added to an oil or gas furnace is usually supplemental and less powerful than the original furnace. If, however, the output of the solid-fuel furnace is greater than that of the system it is added to, then the ductwork may be undersized for the add-on, and heat may not be carried away from the solid-fuel furnace fast enough to prevent occasional overheating. This is good reason to be sure to install an overheat switch in the bonnet of the add-on. Otherwise the heat exchanger of the add-on may be damaged. Undersizing the cold-air return is a common mistake and just as bad as undersizing the leaders (pipes carrying warm air away from the furnace).

CLEARANCES

An important point to appreciate is that clearances between ductwork and combustible materials as recommended by the National Fire Protection Association are greater for wood and coal furnaces than for oil and gas (see **Figure 2.4**). This is partly because solid-fuel furnaces can be operated without electrical power, and

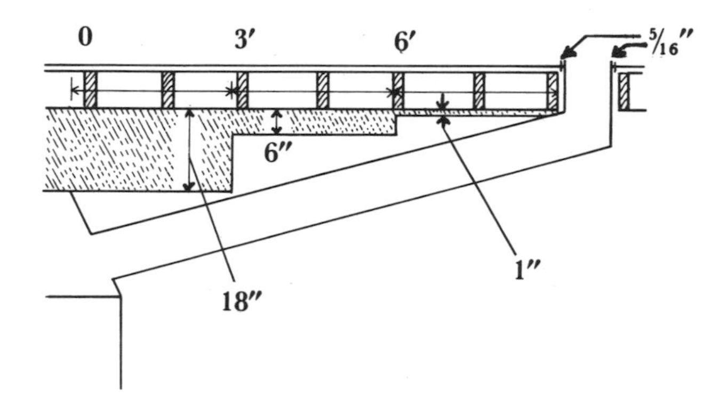

Figure 2.4 Ultimate authority for clearances rests with local or state government. These clearances are consistent with many codes. A good reference is Bulletin 89M of the National Fire Protection Association, 470 Atlantic Avenue, Boston, Massachusetts 02210.

therefore without blowers for hot-air dispersal. Without the blower the ductwork may not be large enough to carry away heat from the furnace by gravity quickly and prevent the furnace from overheating.[5] Another reason is that safety mechanisms in solid-fuel furnaces are comparatively easily overridden, for example, by leaving the ash pit door ajar. This would allow the fire to burn uncontrolled, since a limit switch in the bonnet sensing a temperature of 250°F and shutting the primary damper would have no effect on flow of air into the firebox through the ash pit door. Clearances in Figure 2.4 may be reduced from 18 inches and 6 inches to 9 inches and 2 inches, respectively, when a heat shield of 28-gauge (U.S. Standard Gauge, .0193 inch thick, a little more than 1/64 inch) sheet metal (galvanized steel) is mounted 1 inch from the joists or other combustibles using noncombustible spacers. Porcelain electric fence insulators are excellent for this job. Note that the clearances specified are from duct or bonnet to the combustible material and not to the shield.

A common problem in dual installations is that clearances between ductwork and combustibles, adequate for the original liquid-fuel furnace, are not adequate for the solid-fuel add-on. To observe the clearances cited above, it may be necessary to alter the ductwork connected to the original furnace or to make use of sheet metal heat shields. These are very effective and much better than asbestos board shields. Tacking asbestos board—either asbestos cement board or asbestos millboard—directly to woodwork is not adequate protection. Keeping electrical wiring away from the ductwork is also a point to be observed, since heat will speed deterioration of the insulation surrounding the wires.

The clearances cited above are for hand-fired coal and wood furnaces. The bonnet and first three feet of ductwork from stoker furnaces may be six inches from combustibles as long as there is a limit switch in the bonnet that shuts off the stoker when the bonnet temperature reaches 250°F and—according to the NFPA—a barometric draft control that cannot be set higher than 0.13 inches of water. The reason for the reduced clearance is that the fuel load in the firebox of a stoker is limited to the small amount of fuel actually burning in the firepot. Therefore the potential energy release is small compared with a hand-fired furnace in which the firebox may be stuffed full of coal or wood. The limit switch of the stoker cuts off flow of fuel and air to the firebox just the way it does in an oil or gas furnace.

INSULATION

In basements where space heat is not desired, insulating the ductwork with fiberglass or rock wool easily pays for itself in energy savings. Up to 30 percent of a furnace's heat may be lost to the basement through the ductwork and furnace itself. **Table 2.1** shows the thickness of fiberglass with a payback period of three years as a function of heating degree-days. Preinsulated ducts are available with one or two inches of fiberglass from major manufacturers such as Johns–Manville and Owens–Corning. Under no circumstances should foam or paper-backed fiberglass be used because of the fire hazard.

A word of caution is appropriate here. The binder in fiberglass rolls and boards is organic and therefore combustible. Nevertheless, fiberglass with the resinous binder may be used on furnace ductwork as long as the fiberglass is rated up to 250°F and the bonnet temperature cannot exceed 250°F. This is the case with oil and gas furnaces and also with stoker furnaces. Hand-fired wood and

TABLE 2.1
Recommended Fiberglass Duct Insulation for Unheated Basements and Crawlspaces

2000 heating degree-days (65°F base)	4 inches thick
4000	5
6000	6
8000	6
10000	7

Adapted from: S. R. Petersen, *Retrofitting Existing Housing for Energy Conservation: An Economic Analysis*, Building Science Series #64, U.S. Department of Commerce, National Bureau of Standards, 1974. These figures are based on an assumed fuel cost of 45¢ per therm (100,000 BTU).

coal furnaces may have a limit switch in the bonnet which shuts off combustion air at a bonnet temperature of 250°F, but for the reasons just discussed the bonnet temperature in hand-fired solid-fuel furnaces may on rare occasions go above 250°F. This is especially true of gravity systems in which there is no distribution blower to move hot air upstairs. To play it safe with these furnaces, fiberglass blankets without a binder should be used for duct insulation. All the major manufacturers of fiberglass make these blankets. An example is GlasMat 1200 made by Johns–Manville and rated for 1200°F. Wrapped with aluminum foil this should be safe under all conceivable conditions.

DESIGNING A GRAVITY WARM-AIR SYSTEM

Air warmed by the furnace must somehow move from the furnace to other parts of the house if it is to make the occupants feel warm. Before electrification the distribution of heated air was by natural convection ("gravity flow"), that is, through the tendency of heated (and therefore less dense) air to rise and cooler air to fall under the influence of gravity. After electrification, furnaces were outfitted with blowers. These improved heat transfer at the heat exchanger, allowed the use of smaller ducts, and made central heating possible in ranch houses sprawling from one county to the next. But today we are beginning to ask: Why do a job with more when you can do it with less? Why use brute force when there is a gentler solution? If a house is well-insulated, heat can be distributed quite successfully by gravity. Good insulation is the all-important factor.

That is of course not to say that houses with gravity hot-air systems were well-insulated in the old days. Quite the contrary. And they weren't all that comfortable either. **Figure 2.5** shows a pipeless furnace, the simplest of all gravity warm-air systems. Warm air emerges from the center of a large register located near the middle of the house. Cool air returns to the furnace through the periphery

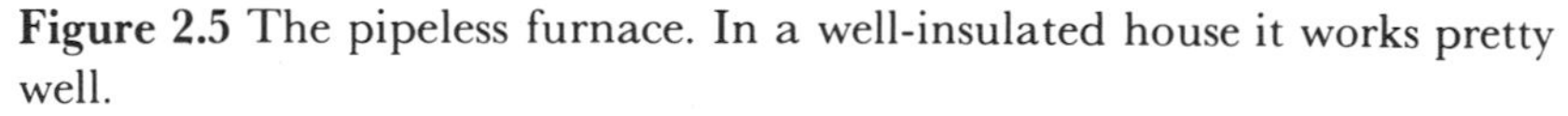

Figure 2.5 The pipeless furnace. In a well-insulated house it works pretty well.

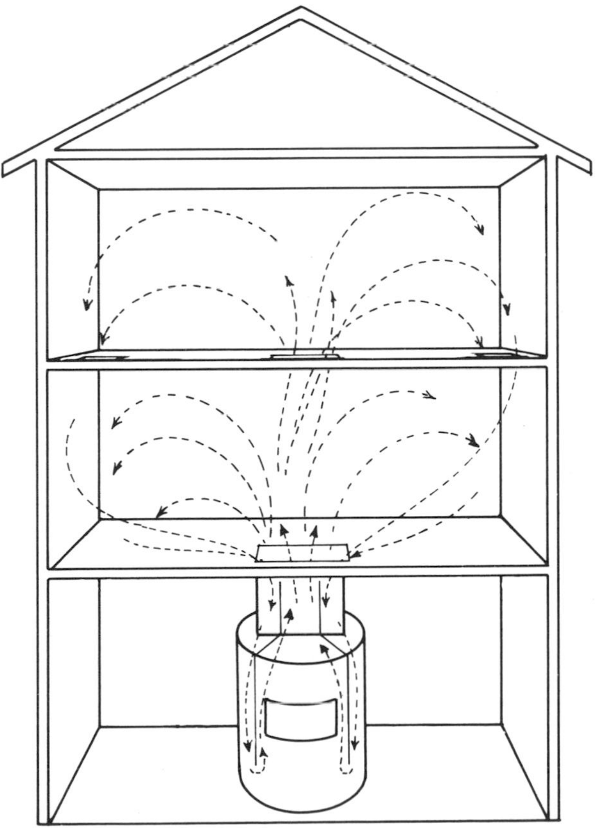

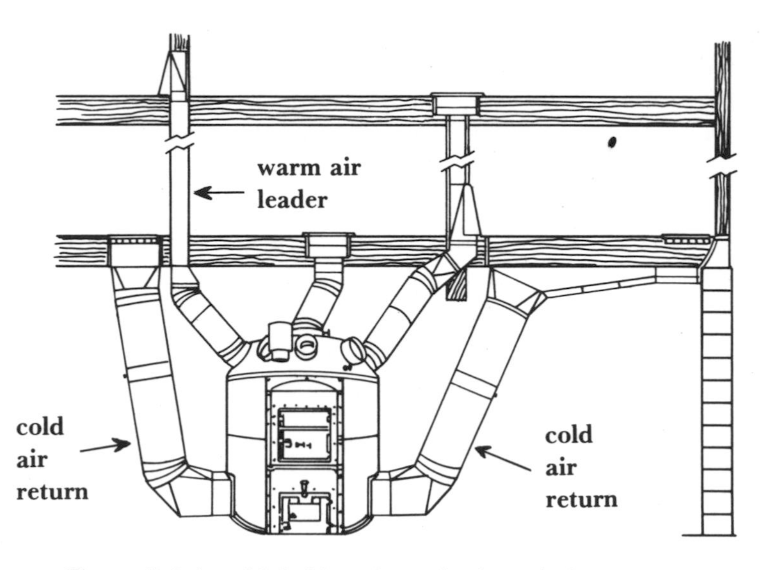

Figure 2.6 An old-fashioned gravity hot-air furnace.

of the same register and descends between an inner and outer shroud in such a way that the returning air is prewarmed. Not bad, only a small heat loss to the basement. However, the house was very unevenly heated. In the middle of the main room, one could imagine himself on a desert island in the South Pacific in the middle of a wind storm. One point in favor of the pipeless furnace is that heat from it is readily dissipated and it is hard to imagine its overheating and causing a house fire.

Figure 2.6 shows the kind of gravity furnace some of us remember from the thirties and forties.[6] This octopus-like monster does a better job of spreading the heat around, but those big cold-air returns don't leave much room in the basement for ping-pong. The house also tended to be dusty because filters would have introduced too much resistance into the system and impeded gravity flow. Thus electricity and the blower brought yet one more benefit into houses heated with coal and wood, namely clean, filtered air. But in new, well-insulated houses the trend can be back toward the pipeless furnace.

Figure 2.7 shows a four bedroom house insulated to today's standard. The walls are packed with six inches of fiberglass (R-19) and the roof is R-32. The heat-loss rate calculated for an outdoor temperature of −20°F is only 18,000 BTU/hr, five to ten times less than what an old-fashioned house would lose. Of course most of the windows are on the south side to make the most of our solar furnace. There is a good argument in favor of a heavy masonry chimney to provide thermal ballast, but the author's preference for insulated stainless steel chimneys is so strong that a light stainless chimney is shown instead. The important point is that a maximum heat-loss rate of 18,000 BTU/hr can easily be supplied by a stove, and not a very large stove at that. The stove could be upstairs, but putting it downstairs keeps dirt and ashes out of the living area. In

Figure 2.7 In a well-insulated house a stove can do the entire heating job.

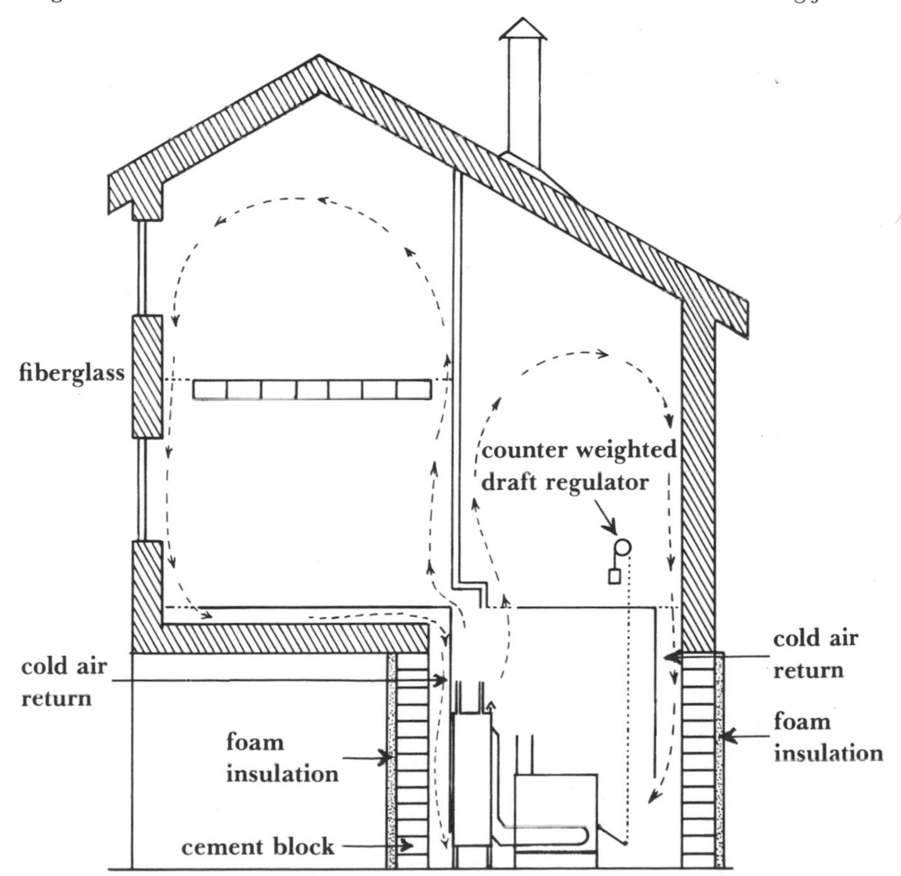

this kind of house money is wisely invested in insulation rather than furnace, ductwork and fuel.

The stove could be surrounded by a well-insulated sheet metal jacket to prevent energy from being radiated directly to uninsulated basement walls. An alternative—and the one shown in Figure 2.7, is to build a fireproof insulated furnace room around the stove. Warm air rises by gravity through two small floor registers which should be double-walled with an air gap of $5/16$ inch between the two walls.[7] These registers should not be of the closeable type, so that there is no possibility of trapping heat in the furnace room. Slightly cooled air comes back to the furnace room through two "cold-air" returns. These can be made very simply just by enclosing part of the space between two joists with a piece of sheet metal; the metal downcomers must extend close to the floor so that there is no chance of hot air rising in them. Should circulation turn out to be inadequate, one could give it a boost by installing a small fan in one or both of the cold-air returns.

The stove room may contain a storage tank for wood- or coal-heated domestic hot water. If the storage tank is left uninsulated, or is only moderately insulated (one inch fiberglass), it provides thermal ballast that can carry a well-insulated house overnight when the outside temperature is 0°F.

Because the furnace room is small and well-insulated, extra care should be taken that all safety rules are rigorously observed. The safest way to build it is of non-combustible materials such as the cement blocks shown in Figure 2.6. Just to be absolutely safe, install an overheat alarm and a smoke detector in the furnace room, and then sleep peacefully.

SMOKE AND HEAT DETECTORS

There are two kinds of smoke detector, photoelectric and ionization. The former is more responsive to dense smoke from smoldering fires and the latter more responsive to hotter fires that give off smaller particles.[8] Photoelectric detectors incorporate a small photocell that generates a small potential difference when light scattered by smoke particles falls on it. In an ionization detector a small amount of radioactive material gives off alpha particles that break up smoke particles into ions, which migrate toward charged electrodes and cause a current to flow in an alarm circuit.

Since it is hard to predict which kind of detector will be most appropriate when a fire actually does start, it is hard to make a choice on their differences in performance alone. However, those who question the wisdom of using radioactive americium-241 in detectors will choose the photoelectric kind. True, the americium presents no hazard as long as it is tightly sealed inside the detector, but one can imagine various ways it can get out. Theoretically all ionization detectors should be turned in to the Nuclear Regulatory Commission for safe disposal when their useful life is at an end.

Placement of detectors is crucial. If a chimney fire is what you consider to be the most likely hazard, then a heat detector attached to the chimney in the attic may give you earliest warning. Its alarm should be mounted in or near the bedrooms. A complete early warning system includes a heat detector near the furnace as well as a smoke detector located between the bedrooms and the rest of the house.

USE OF OUTSIDE COMBUSTION AIR

Generally speaking, ducting combustion air from the outside directly to the firebox of a furnace is neither necessary nor desirable. Tight houses are usually also well insulated, and therefore the heating load low. In the house in the example above (Figure 2.7) the amount of air required to produce 18,000 BTU/hr is on the order of $\frac{1}{16}$ of a complete air change per hour. This is below the air exchange rate at which the house begins to smell stale. Ducting outside air to the firebox would uncouple the chimney from the living area and thus worsen the odor problem.

Turning to larger furnaces in drafty houses, conducting outside air directly to the firebox can improve the performance of the furnace if the chimney draft is reduced by competing drafts in the house. In houses poorly sealed near the top, the entire house can act as a phantom chimney pulling on the same inside air as the furnace and its chimney. This makes the furnace smoke and respond poorly. Plugging the leaks in the house is probably a more effective way of solving this problem than providing the furnace with outside air, since this reduces unnecessary heat-loss from the house, improves furnace performance, and at the same time helps to humidify the house.

It is sometimes claimed that use of outside air for combustion increases the thermal efficiency of the house, the thought being that this stems loss of heated air to the furnace. This argument is even less compelling for a furnace in an unheated basement than for an upstairs stove. The counter-argument is that introduction of outside air into the firebox lowers temperatures throughout the furnace-flue system with the possible results being: reduction of heat transfer efficiency in the heat exchanger, reduced combustion efficiency, and increased deposition of creosote in the chimney. Lacking experimental evidence, it seems wisest to use inside air for combustion unless the furnace is so consistently smoky that you are willing to try any remedy. Be sure to connect the duct bringing air to the firebox to the side of the house against which prevailing winds blow and where the air pressure is normally highest.[9]

NOTES TO CHAPTER II

1. Metlvent, Metalbestos and Pro-Jet are brand names of insulated stainless steel chimneys.
2. L. B. Schmitt and R. B. Engdahl, "Performance of Residential Chimneys", *ASHVE Transactions,* 1949, page 241.
3. P. Nicholls and C. W. Staples, "Removal of Soot from Furnaces and their Flues by the Use of Salts or Compounds", *Bulletin 360*, U.S. Department of Commerce (Bureau of Mines) 1932.
4. The positions of the two furnaces may be interchanged if the distribution blower is removed from the oil or gas furnace and repositioned upstream of the solid-fuel furnace.
5. If the power does go off, gravity circulation may be greatly improved by removing any air filters in the system, opening a door to the basement, and by opening the door to the blower compartment of the furnace.
6. Standard procedures for sizing gravity warm-air systems of the kind shown in Figure 2.6 were worked out in the thirties, but are not easy to find today. They are also not very relevant to well-insulated houses. For those interested, procedures may be found in (a) the ASHRAE *Guide,* 1948 and years preceding; or (b) James D. Hoffman, ed., *Gravity Warm*

Air Heating—Digest of Research, National Warm Air Heating and Air-Conditioning Association, Columbus, Ohio, 1935.
7. See Hoffman, above page 507.
8. In one series of tests on smoke from a urethane mattress, five out of nine ionization detectors failed to respond at all, whereas all three photoelectric detectors responded to fairly low levels of smoke. Walter J. Schuchard, "Smoldering Smoke", *Fire Journal,* January 1979, page 27.
9. For a thorough treatment of the question of outside combustion air, see J. W. Shelton, in *Solar Age*, September 1979, page 30.

Some new evidence has come to light which throws into question the ability of stainless steel chimneys to resist attack by flue gases from coal. It is still too early to tell whether this means it is unsafe to use stainless steel chimneys for venting gases from coal fires. Until this question is answered, one should proceed cautiously and try to stay informed.

CHAPTER III

Solid-Fuel Boilers

Boilers are not essentially different from hot-air furnaces, except that the heat is distributed by hot water or steam rather than hot air. The basic features of full-fledged boilers are the same as those of furnaces: firebox, heat exchanger, and a circulator, which for a boiler is a water pump instead of a blower. Hot-water heating systems are initially more expensive than hot-air systems, but they have several advantages: room temperature tends to be more constant; heat is more cheaply and conveniently transported over long distances in water than in air; boilers are quieter and less dusty than furnaces.

HOT-WATER AND STEAM SYSTEMS

"Boiler" is a misnomer when applied to a hot-water heating system for the water in such a system doesn't really boil. The term is a carryover from the days when steam boilers were high technology. It was the most natural thing in the world to take that technology out of factories and put it in residential basements to provide warmth instead of mechanical power. But steam boilers are now on their way to obsolescence for residential heating, because hot-water systems are quicker to respond, give more even heat, suffer less from corrosion, require smaller pipes; and because they don't whistle, hiss, gurgle, and bang. A major disadvantage of hot-water systems is that they can be badly damaged if allowed to freeze.

Although all the boilers discussed in this chapter are intended for hot-water ("hydronic") heating systems, there is no essential difference between boilers used with hot water and boilers used with steam. The boiler is completely filled with water in a hot-water heating system and only partially filled in a steam heating system. Steam boilers must have a sight glass for determining the water level in the boiler, and must also have one or two low-water cutoff mechanisms whose purpose is to prevent tubes in the top of the boiler from burning out if the water level gets too low.

HOT-WATER SYSTEM BOILERS

The majority of boilers on the market today are made of steel. An exception is the cast iron Tasso shown in **Figure 3.1,** an import from Denmark which burns either wood or coal. A drawback to cast iron is that castings can crack, especially large boiler castings. When a casting breaks, the entire boiler must be taken apart and the broken casting replaced. This can be expensive. Manufacturers tend to prefer steel because it is comparatively easy to work with.

Automatic control of a solid-fuel boiler is significantly different from control of a solid-fuel furnace. The boiler's primary damper or combustion blower is activated by an electrical or mechanical thermostat immersed in the boiler water, not by an upstairs room thermostat. Such an immersed thermostat is called an "aquastat". If the boiler temperature falls below some setting, usually around 180°F, air is admitted to the firebox until the water temperature rises back up to operating temperature. When the upstairs thermostat calls for heat, a circulator pumps 180° water through radiators in the house. A big advantage of a boiler compared to a furnace is that it can provide domestic hot water at a constant temperature. This is done through a heat exchanger immersed either directly in the boiler's water reservoir or built into an external loop.

Overheating of a hot-air furnace is prevented by a high-limit switch in the bonnet that automatically cuts off flow of air to the firebox if the bonnet temperature exceeds 250°F. With boilers, if the water exceeds some high-temperature limit, usually around 200°F, the circulator automatically goes on and heats the house regardless of house temperature. This is universally referred to as "heat dumping".

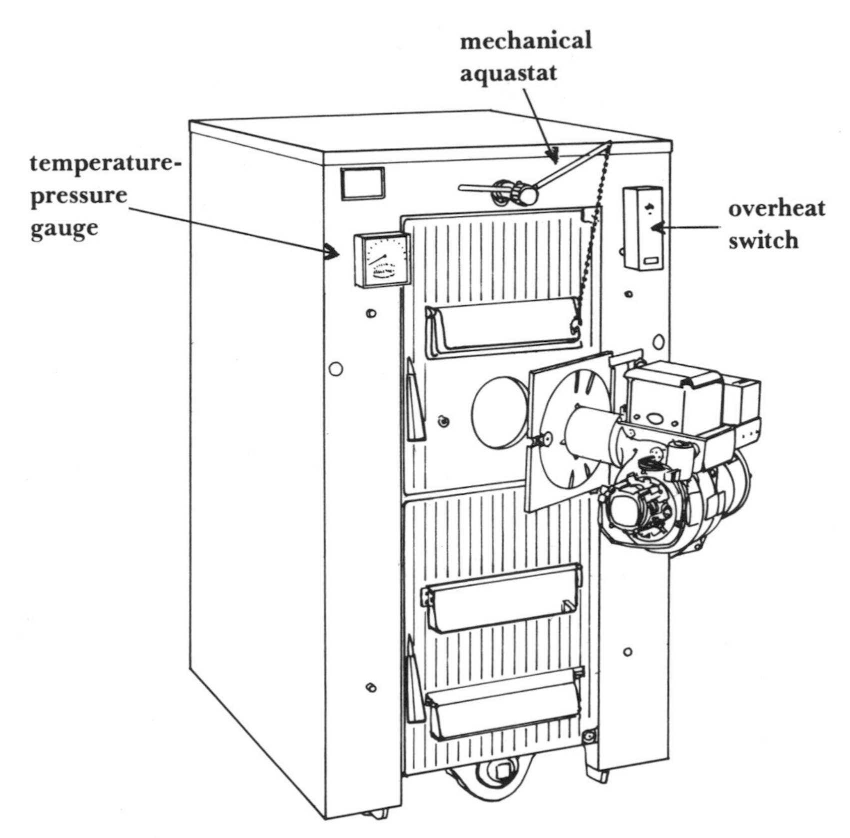

Figure 3.1 The Tasso Universal boiler. It burns wood, coal, or oil. The oil gun is mounted on a hinged door and can be swung out of the way when solid fuel is burned. Like the other European imports, all Tasso models come with a mechanical aquastat. The Tasso Company has been in business 94 years.

SELECTING A BOILER

Some of the important features to look for in a hot-water boiler are the same as those for a hot-air furnace, but there are special complexities in hydronic systems.

Water Volume. The more water the better. This conclusion follows from the fact that a solid-fuel fire cannot be cut off instantaneously, as can an oil or gas fire. When an oil or gas boiler reaches temperature, the aquastat shuts off the flow of oil or gas to the burner and the fire immediately goes out, but in a solid-fuel boiler, it is the flow of air to the firebox that is cut off. The fire does not go out abruptly, but rather coasts down gradually. During coast-down heat is still given off and the temperature of the boiler water may continue to rise. The less water the boiler holds, the greater the temperature overshoot and the greater the likelihood that heat will be dumped in the house. This is a big nuisance. Avoid it by getting a boiler that can hold plenty of water and soak up the extra heat without a very big temperature overshoot.

Generally speaking, solid-fuel boilers cannot produce heat as fast as the oil or gas boilers they replace. This, too, argues in favor of a larger volume of water so that extra heat stored in the water can make up for the slower heat-generating capability of the solid-fuel boiler. To put it another way, extra heat storage capacity should be built into a solid-fuel boiler to compensate for lower power and less precise control. The only real advantages of a boiler with a small water capacity are lower initial cost and faster response from a cold start.

When a solid-fuel boiler is connected in parallel with an oil or gas boiler, most of the water volume can be in the oil or gas boiler. **Figure 3.2** shows a Thermo–Control wood stove supplying heat to a conventional boiler. Even though the coils in the firebox of the Thermo–Control contain only about one gallon of water them-

Figure 3.2 The Thermo-Control. A significant fraction of the fire's heat is transferred to surrounding air. The Thermo-Control is an add-on and intended to assist a full-fledged boiler.

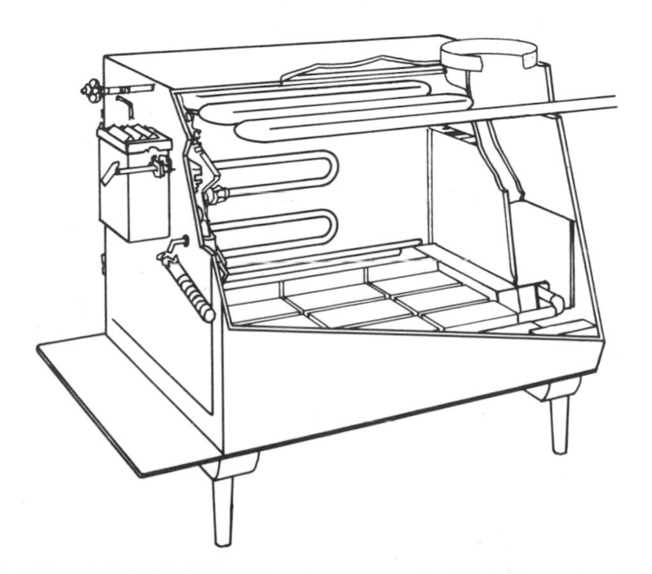

selves, the entire system may contain a good deal of water and enough heat-storage capacity to prevent wild temperature excursions.

Configuration of the Boiler. After the stove with a coil in the firebox, the next step up in complexity is a boiler in which water is contained between two concentric cylinders. The Passat (**Figure 3.3**) is an example. The basic model lacks enough surface area between firebox and water jacket for good heat transfer, so the Passat comes in two different forms—with and without a drum-like heat exchanger attached to the rear end. You might think that the more efficient model would be the obvious choice in all cases, but here,

Figure 3.3 The Passat is made of two concentric cylinders. There are two loading doors: one for ordinary chunks mounted on a front that swings away entirely in case you don't like splitting wood. That's an oil burner piggybacking the main firebox.

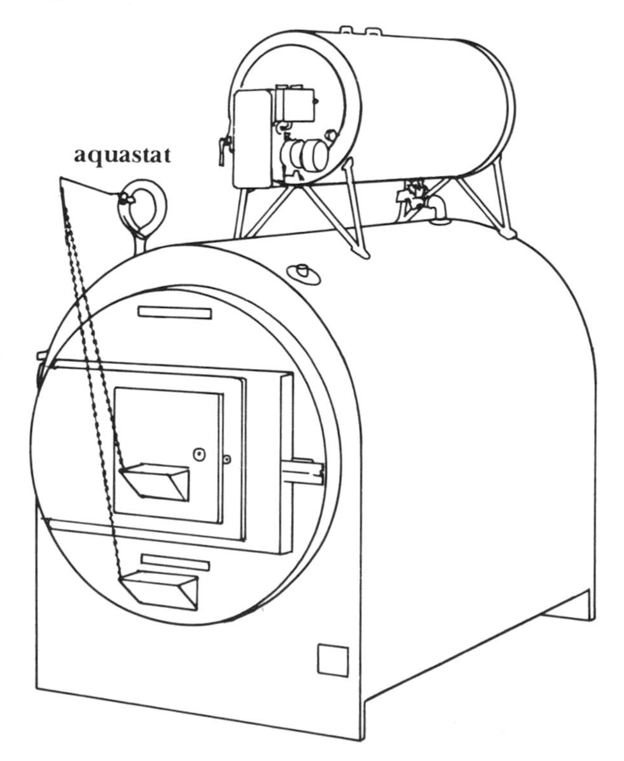

again, one must weigh efficiency against safety. It is up to the heating contractor and buyer to choose the appropriate model for a given setting. Where the flue passage is long and the smoke subject to excessive cooling, a model without the heat exchange drum is more appropriate if minimization of creosote deposition is of primary importance.

The Passat Company is Denmark's next-to-largest boiler manufacturer. Largest is the Tarm Company, whose MB–Solo is shown in **Figure 3.4.** Here firebox size is sacrificed for lots of indirect heat-exchange surface, that is, surface warmed by flue gases but not by direct radiation from the fire. Thus the Passat and Tarm are based on two different philosophies, one with a large firebox for conven-

Figure 3.4 The Tarm MB–Solo with a grate for either wood or coal. The coil at the top supplies domestic hot water.

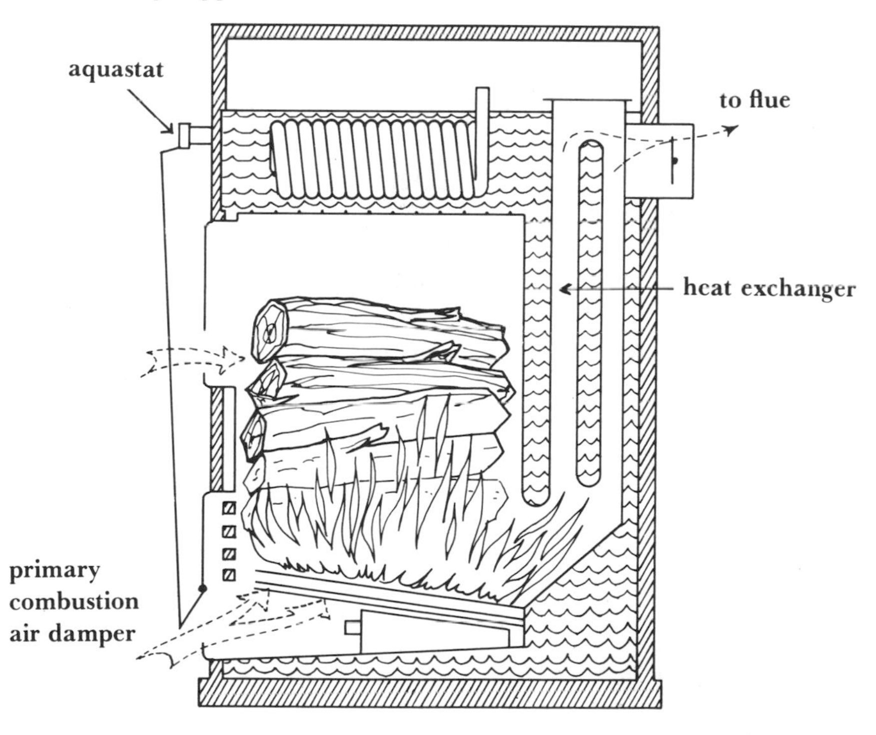

ience and one with lots of heat-exchange surface for high heat-transfer efficiency. Both brands have enthusiastic supporters. A penalty the Tarm and similar boilers pay for good heat exchange is a relatively short burn as a consequence of a small firebox. The Passat, with its much greater wood capacity and grateless firebox, burns bigger wood longer. The entire front of the Passat can be opened in case you want to get rid of an old chair quickly. Tarm boilers are built for coal as well as wood. The Passat, lacking grates, is intended for wood only.

Thickness of Steel. The Boiler Code of the American Society of Mechanical Engineers (ASME) calls for a minimum steel thickness of 1/4-inch for all heat-exchange surfaces in low-pressure boilers.[1] In most localities the ASME Boiler Code standards have been incorporated into codes covering safety in public buildings, so its provisions are not binding on manufacturers who supply residential boilers. Nevertheless, most manufacturers try to meet the Boiler Code so that their boilers can be installed in public buildings.

As an aside, the inner cylinder of Passat boilers intended for the European market are made of 3/16 inch corrosion-resistant steel, but since the ASME Boiler Code gives no credit for use of high-grade steel, Passats exported to the United States are made of 1/4-inch common steel. This might be one of those instances where a code developed to insure high quality inhibits innovation and even improvement.

Aquastat. The mechanical aquastat shown on the boilers in Figures 3.1 and 3.4 is constructed of a bimetal helix in an immersion well connected to a rod and chain. As the temperature changes, the bimetal helix winds tighter or looser and the rod is rotated, thus lifting or lowering the air-controlling flap. Mechanical aquastats have the great advantage over electrical that they continue to function even when the power is off. They are cheaper than motorized dampers, rugged and quiet.

Insulation. The best-insulated boilers today are by Passat and Hoval with 2 inches of mineral wool. Up to 6 inches of mineral wool can be justified where heat lost to the basement is considered wasted. Most house doors are on the order of 28 to 30 inches wide, and this limits the amount of factory-installed insulation on equipment that must pass through them. It is a fair conclusion that most boilers today are under-insulated.

A NEW IDEA

Most of the furnaces and boilers new to the market, although entirely serviceable, are not much different in basic design from the ones of 100 years ago, except that control is more sophisticated today. Professor R. C. Hill, however, working under a Department of Energy grant at the University of Maine, has been rethinking the whole problem of burning wood and doing something about it.

Hill's idea is to burn wood as efficiently as possible by applying modern engineering techniques to residential boilers. His boiler includes an insulated refractory combustion chamber where the burning temperature is kept around 1500°F and air is injected into the combustion region under high pressure to promote good mixing of oxygen and wood gases. These two features ensure nearly 100 percent combustion efficiency. The other unique feature of Hill's boiler is that it is coupled to a high-capacity heat-storage water tank—on the order of 500 gallons (**Figure 3.5**). This permits long full-blast firings—on the order of 2 to 6 hours—once a day and eliminates the smoldering phase of the burning cycle when so much creosote is generated in a conventional boiler.

Between the combustion chamber and heat-storage tank is a more-or-less standard heat exchanger. Because the burning rate is always the same, the heat-exchange surface can be matched to it for optimum efficiency without compromises for the sake of minimizing creosote deposition. Overall efficiency of the system is on the order of 75 percent vs. 55–60 percent for conventional boilers. No creosote; no chimney fires.[2]

The main drawback to Hill's way of burning wood is the very long recovery period if the system ever is allowed to cool below, say, 100°F. It takes a long time to heat 500 gallons of water. Nevertheless, Hill clearly has an idea whose time has come. He has taken his boiler through several stages of development and believes it is now ready for general residential use. There are three companies making Hill-type boilers, one of whose products is shown in **Figure 3.6.**[3] Two or three logs are placed into the cylindrical combustion tube of

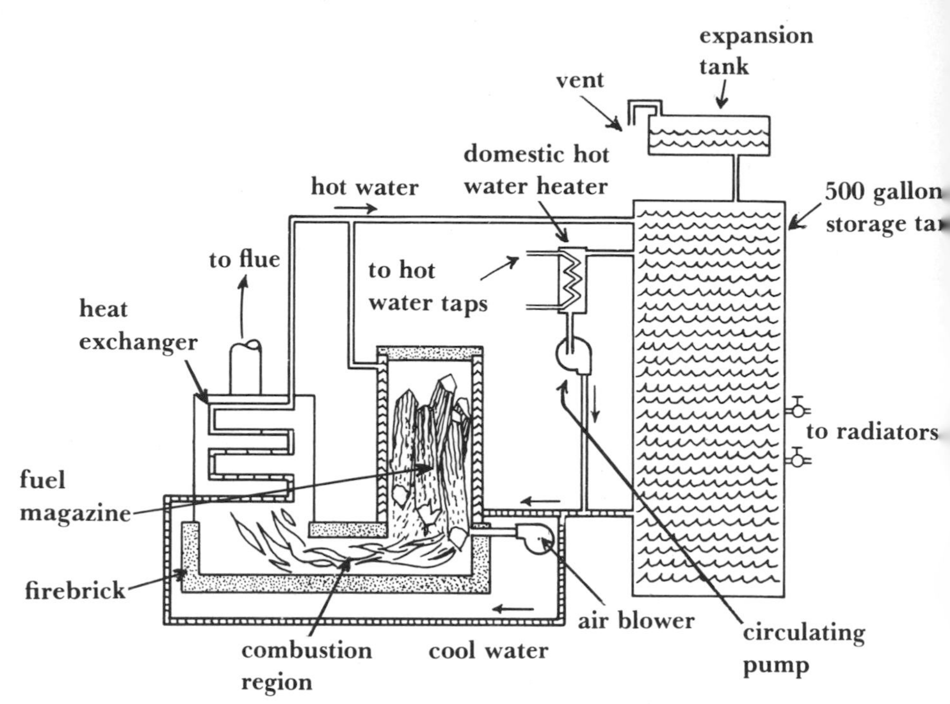

Figure 3.5 A diagram of one of R. C. Hill's test models. Wood is loaded vertically into a cylindrical steel fuel magazine. Logs gradually settle into the combustion region as their lower ends burn away. The temperature in the combustion region is maintained at about 1500°F. Heat from the exhaust gases is transferred to a large tank of water from which it is distributed to the house as needed.

the Jetstream from above and burning takes place in a base made of precast refractory cement. A jet of air that enters through a stainless steel tuyere impinges on the glowing wood with such force that it blows away all ash on the surface of the wood so that oxygen comes in direct contact with the surface. Any incompletely burned gases are further oxidized in the red-hot refractory tunnel before reaching the heat exchanger—an array of 16 one-inch tubes.

Water heated in the boiler is transferred to a storage tank whose size depends on the heat demand of the house and the owner's schedule. The Jetstream is rated at 120,000 BTU/hr by the manufacturer, and is normally coupled to one or more storage tanks with a total capacity of 400–600 gallons. This is enough to carry a reasonably well-insulated house for one or several days—depending on outside temperature—at a maintenance temperature of 40°F if the initial tank temperature is 180°F. One owner who was determined to fire only on Saturdays installed 2200 gallons of storage capacity in his basement in the form of eight 275-gallon oil tanks.

Storage tanks can be made of steel or fiberglass or wood. A heat exchanger for domestic hot water can be immersed in the storage tank or installed in the loop between boiler and storage. Tanks rated to withstand conventional boiler pressures of up to 30 pounds per square inch (psi) would be quite expensive, and therefore the system is normally left open to the atmosphere, communicating with the air through an expansion tank which must be at the top of the system to keep water from spilling.

Heat-storage is an indispensable component of Hill's design, but when you think about it, you realize that heat storage could also be used to advantage with conventional boilers. By uncoupling heat production from heat demand, smoldering—and therefore creosote—can be avoided with almost any boiler. Not a bad idea. Why didn't we think of it earlier? Heat storage also makes sense in commercial buildings which are left unattended over weekends, or in restaurants and motels where there are very pronounced peaks in demand for domestic hot water at certain times of day.

Of course a Hill-type boiler and associated plumbing are more

Figure 3.6 The Jetstream by Hampton Technologies is the same in all essentials as the prototype of Figure 3.5. Both heat exchanger and fuel magazine are incorporated in the same water jacket. The refractory base and steel boiler are shipped separately and put together on site. The storage tanks in this illustration are prefabricated from fiberglass.

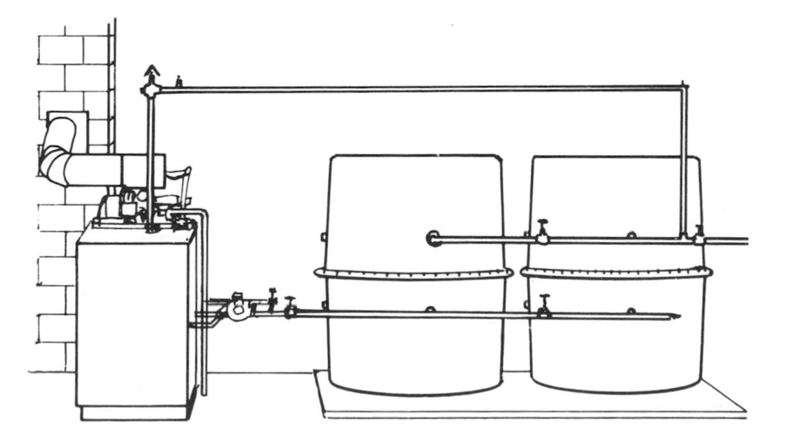

TABLE 3.1
Savings Schedule for Conventional and Hill-Type Storage Boilers
(1980 dollars)

	Price of Oil[a]	*Cost of Oil*[b]	*Price of Wood*[c]	*Cost of Wood*[d]	*Fuel Saving*	*Conventional Boiler Accumulated Saving (55%)*	*Hill Boiler Accumulated Saving (75%)*[e]
80–81	$1.20	$1200	$75/cord	$442	$758	−$2500 −1769	−$4400 −3525
81–82	1.284	1284	77.25	455	829	−966	−2576
82–83	1.373	1373	79.56	469	904	−90	−1548
83–84	1.470	1470	81.95	483	987	868	−433
84–85	1.572	1572	84.41	498	1074	1913	733
85–86	1.683	1683	86.94	512	1171	3053	2079
86–87	1.800	1800	89.55	528	1272	4294	3491
87–88	1.926	1926	92.24	544	1382	5644	5017
88–89	2.061	2061	95.00	560	1500	7112	6666
89–90	2.206	2206	97.85	577	1629	8707	8448

[a] Escalation rate = 7% per year above general inflation rate
[b] 1000 gallons
[c] Escalation rate = 3% per year above general inflation rate
[d] 169 gallons of oil equivalent to 1 cord good hardwood; oil boiler 65% efficient, wood boiler 55% efficient
[e] Same as preceding column, except wood boiler is 75% efficient, i.e., 1 cord = 230 gallons oil.

expensive than a conventional boiler without heat storage. But it should be remembered that the extra efficiency of the Hill boiler saves extra gallons of oil or cubic feet of gas and therefore results in considerably higher fuel savings. An economic comparison of Hill and conventional boilers is made in **Table 3.1** assuming initial investments of $4,400 and $2,500, respectively. At the end of ten years the payback is about the same under our assumptions. If the price spread between oil and fuelwood grew faster than assumed in Table 3.1, the Hill boiler would have a clear economic advantage.

Although the Hill boiler uses water as the primary heat-conveying medium, the boiler can be adapted to hot-air systems simply by installing a water-to-air heat exchanger in the ductwork. In a similar way, the boiler can be used with a rock bed for storage instead of a water tank.

LARGE BOILERS

Most solid-fuel boilers on the market today are intended for residential use; most are rated at less than 200,000 BTU/hr. Riteway boilers range up to 300,000 BTU/hr and the biggest Passat is rated at 600,000 BTU/hr by the manufacturer. On the other end of the scale, steam-power boilers bigger than 2 million BTU/hr are readily available.[4] Anyone shopping for a hot-water boiler in the intermediate range will find that there is a small selection, although industry will doubtless fill this gap in the next few years. Where a large oil or gas boiler is already in place, the best short-run solution may be to convert to wood or coal by adding an external ("dutch-oven") firebox for wood or rebuilding the original firebox for coal. Fluid-fuel fireboxes are almost always too small to accommodate a reasonable amount of wood. If a dutch-oven is added, care must be taken to minimize air leakage at joints between refractory materials and steel. Furthermore, there must be easy access to heat exchange tubes for cleaning.[5]

NOTES TO CHAPTER III

1. Section IV of the Boiler Code treats low-pressure residential boilers.
2. A technical report on the Hill boiler may be obtained by sending $2.00 to the Department of Industrial Cooperation, University of Maine, Orono, Maine 04469. Make the check payable to the University of Maine.
3. Dumont Industries, Monmouth, Maine 04259
   Madawaska Wood Furnace Company, 86 Central Street, Bangor, Maine 04401
   Hampton Technologies, Ltd., Box 2277, 126 Richmond Street, Charlottetown, Prince Edward Island, Canada C1A 8B9
4. A listing of companies that manufacture large wood- and coal-fired steam boilers is available: John Bressoud, "A User's Guide to Wood Engineering and Equipment". DSD #135, Resource Policy Center, Thayer School of Engineering, Dartmouth College, Hanover, New Hampshire 03755.
5. Plans for adding a dutch-oven to a gas or oil furnace or boiler are available from Bill White, The Firebuilders, 352 Stetson Road, Brooklyn, Connecticut 06234.

CHAPTER IV

Installing Hot-Water Boilers

Hooking up a boiler is more problematical than connecting a furnace to ductwork. There are more things to go wrong and satisfactory performance is harder to obtain, especially when the solid-fuel boiler is added to a gas or oil boiler. The situation is complicated by the fact that oil and gas boilers seldom have fittings for making connections to the solid-fuel boiler in just the right places. The operator must interact with a solid-fuel boiler in a way never required by an oil or gas boiler.

A boiler is a potential bomb. Except near the freezing point, water expands when heated in an open container. If it is heated in a closed container so that expansion is prevented, its pressure rises dramatically. Theoretically, the pressure in a residential boiler would rise by more than a thousand-fold during warmup, if expansion were not allowed. The boiler would burst long before reaching its operating temperature. Fortunately, safeguards are so well worked out, and plumbers are so familiar with them, that accidents are rare today. But every now and then one reads of workmen in a power plant being scalded by steam when something goes wrong with a high-pressure boiler. Residential boilers are low-pressure boilers, operated either open to the atmosphere or as part of a closed system in which the pressure is normally kept at about 15 psi above atmospheric pressure through use of an expansion tank not communicating with the atmosphere and partially filled with air. This limit of 15 psi goes back to the old days when residential boilers and radiators were made of cast iron, which has a relatively low tensile strength. Well-engineered steel boilers can easily be operated at much higher pressures. Modern steel power boilers, for example, are operated at pressures on the order of 1000 psi.

OPEN BOILER SYSTEMS

Open systems are referred to as "unpressurized", although the pressure of the water in the boiler itself is in fact somewhat elevated due to the weight of water in the pipes above it. In buildings taller than thirty feet this hydrostatic pressure can be more than the 15 psi common in closed residential systems. Open systems were the norm in the United States in the old days, and still are found extensively in Europe. Almost all new residential systems in the United States are closed, however, to avoid having to put the expansion tank at the highest point in the system, namely the attic where it may be subject to freezing.

SAFETY IN CLOSED (PRESSURIZED) BOILERS

In a pressurized system (**Figure 4.1**) expansion takes place at the expense of an air cushion trapped in the expansion tank. Modern closed expansion tanks have a flexible diaphragm between water and air. The pressure on the system can be adjusted by varying the air pressure in the tank through a pneumatic valve like those used to inflate rubber tubes and tires. When the water is heated and expands the pressure on the system does not rise much because the air cushion is easily compressed.

An important point to be aware of is that the total volume of the system, including pipes and radiators, must be taken into account when sizing the expansion tank. If the tank is too small, small temperature increases will cause a rise in pressure sufficient to open the pressure relief valve. This is to be avoided, since cold makeup water flowing into the system contains oxygen, which causes early deterioration of the system through corrosion. (See Appendix 2 for a method of sizing expansion tanks.)

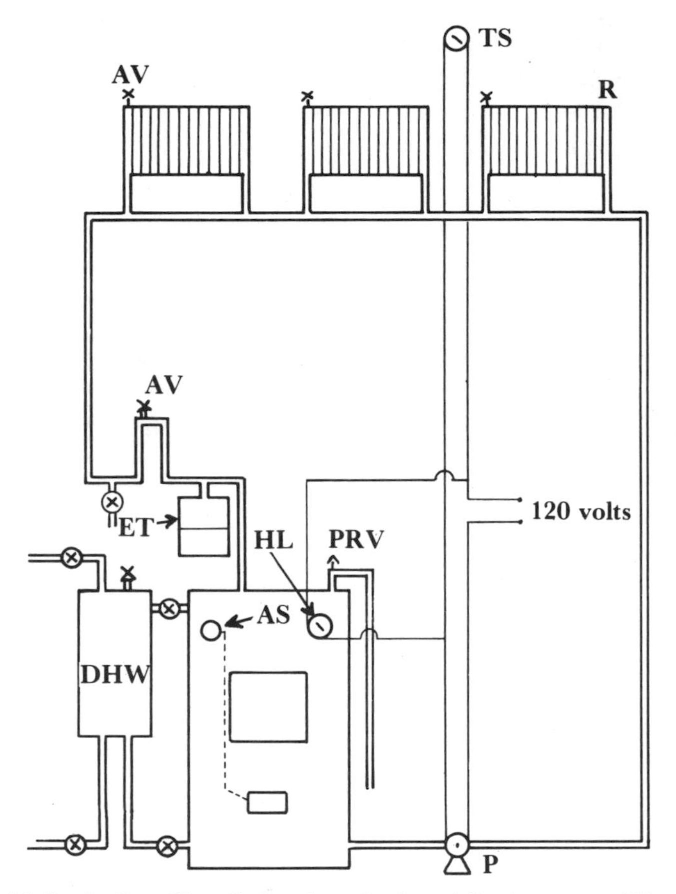

Figure 4.1 Basic One-Pipe Boiler Installation. AS—Aquastat; HL—High-limit or overheat switch; ET—Expansion tank; AV—Air vent; TS—Thermostat (line voltage); R—Radiator; P—Circulator pump; DHW—Domestic hot water. PRV—Pressure relief valve.

A spring-loaded pressure relief valve is set to go off at 30 psi. Low-pressure boilers, constructed according to the ASME Boiler Code, are all shop-tested to 60 psi, so there is quite a margin of safety. If you look at the tag on a pressure relief valve, you will find that it is rated not only with respect to pressure, but also with respect to power in BTU/hr. This power rating tells how fast hot water—and therefore energy—can be discharged through the valve orifice when the pressure difference across the valve is equal to the valve's pressure limit. The power rating of the valve should exceed the maximum conceivable heat input rate to the boiler under runaway conditions. It would be a serious mistake to use a pressure relief valve with an inadequate power rating or to reduce its capacity to dump heat by reducing the size of the discharge opening.

Because the expansion tank and the pressure relief valve are so vital to the safe operation of a boiler, they should be mounted close to or right on the boiler in a pressurized system and never at a position in the line where turning a valve could isolate them from the boiler.

STANDARD BOILER INSTALLATION

Figure 4.1 shows a typical boiler installation. An aquastat opens the air shutter to stimulate the fire whenever the water temperature falls below the desired level. The circulator pump is activated by either the upstairs thermostat or the overheat switch at the top of the boiler. Air vents, preferably automatic, are absolutely essential at the very top of the system and anywhere else that air bubbles might become trapped and stop water flow or reduce the heating effect of the radiators. The sophisticated plumber will slope pipes in such a way that direction of circulation and gravity help air bubbles to flow toward the air vents. Saddle-type air vents may be retrofitted on copper tubing where needed without opening the system. However, they are definitely second-best compared with automatic air vents.

Heat-Dumping. As mentioned earlier, solid-fuel boilers are prone to overheat, especially in mild weather. Many people who have installed solid-fuel boilers have discovered too late that they or the plumber should have provided for heat-dumping to avoid blow-off. A temperature-sensitive switch can be retrofitted on one of the hot pipes on the top of the boiler to take care of this problem without draining the boiler. An example of such a switch is Honeywell LA 409B 1071, an inexpensive mercury switch which closes whenever the temperature rises above a variable limit.

The Pump. The position of the circulator pump happens to be in the cold return in Figure 4.1. If it were in the hot flow line, it might become air-bound, since bubbles are more likely to form there. The disadvantage to locating the pump in the cold return is that it may reduce the pressure at the top of the return riser enough to prevent venting. This is not apt to be a problem with residential low-head pumps.

Should a leak develop or the pressure relief valve open, makeup water is added to the system through a makeup valve which opens automatically when the pressure difference between the system and the atmosphere falls below 12 psi.

Domestic Hot Water. Domestic hot water (DHW) may be provided by a tank-type heat exchanger immersed right in the boiler or by a finned coil, either internal or external to the boiler. In Figure 4.1 boiler water circulates by gravity through an external heat exchanger. The external loop must be well insulated to prevent excessive heat loss to the basement.

The Expansion Tank. Note that the expansion tank (with diaphragm) hangs from the flow pipe in Figure 4.1. If the expansion tank were above the pipe the natural buoyancy of hot water would carry heat up into the tank and warm the air cushion. This is to be avoided because the pressure exerted by the air cushion on the water in the system depends on how hot the water is. By keeping the air cushion cool, the expansion tank has a greater capacity to absorb expansion of the water without reaching blow-off pressure. A diaphragmless expansion tank cannot be suspended or the air cushion would be lost.

Radiators. Flow of hot water is diverted to the radiators[1] through Venturi tee fittings (**Figure 4.2**). With the system shown in Figure 4.1 each successive radiator is cooler than the one before. In large buildings with many radiators this is unsatisfactory, especially for those living in the apartment at the far end of the line. Thus, for large buildings the two-pipe system of **Figure 4.3** is employed. Here each radiator receives water at nearly the same temperature and cooled return water is fed into a separate pipe. The one-pipe system is suitable only for small, well-insulated buildings.

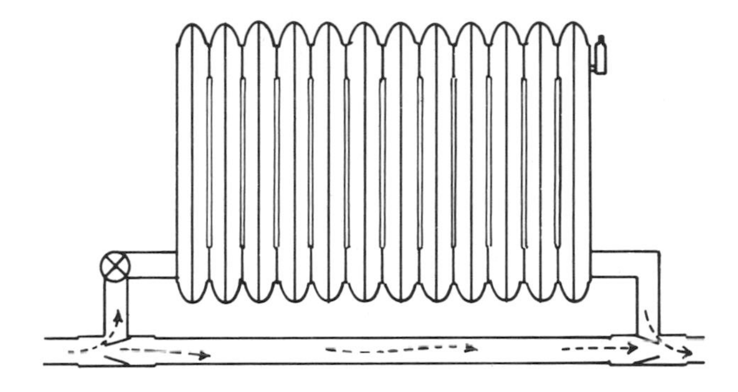

Figure 4.2 Venturi tees direct water into radiators. The one on the downstream side acts as a small jet pump, pulling water through the radiator. One may be used alone, but the effect is doubled by using two.

Figure 4.3 In the two-pipe system water entering each radiator is at nearly the same temperature, but the total length of pipe in the system is greater than in the one-pipe system (cf. Figure 4.1). The expansion tank is connected in the English fashion, i.e., with separate vent pipe and the expansion tank itself connected to the coolest point. If boiling should occur, steam can easily escape through the separate vent pipe. Standard practice in America is to use a single pipe for the expansion tank (cf. Figure 5.6).

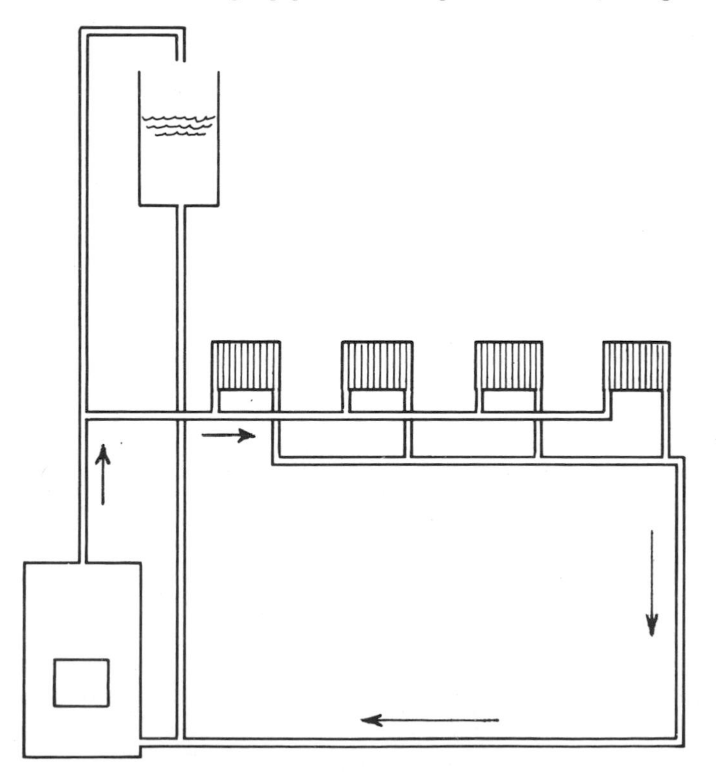

Pipes. Whenever a solid-fuel boiler is added to a gas or oil boiler, the radiators and pipes are already in place and you take what you get. The size of the pipes is not critical in systems where water is forced through the pipes by a pump, because the pump overcomes resistance in the system by brute force. Generally speaking, the size of flow and return fittings may be taken as the manufacturer's recommendation of pipe size. However, for gravity systems, fittings that come with the boiler may be smaller than the ideal size. In residential installations $1\frac{1}{4}$-inch flow and return pipes are more-or-less the rule for forced flow. These are big enough to give some gravity flow during a power outage but not large enough for adequate gravity flow in most cases.

Sizing gravity systems is nearly a lost art and directions on how to do it are hard to find. Therefore, a method for sizing pipes in a gravity system is given in Appendix 3. Adequate circulation by gravity is especially easy to obtain in small, compact, well-insulated houses. If you set up a gravity system and find that heat delivery is not sufficient, a pump can always be spliced into the line.

Unless you intend to heat the basement, it makes sense to insulate the boiler pipes. Most basement walls are completely uninsulated, and heat-loss through them is excessive. The R-value of an 8-inch concrete wall is about 1, compared to R-24 or more for 8 inches of fiberglass. In the olden days boiler pipes were insulated, if at all, with corrugated asbestos paper (R = 1 per inch). Until not too long ago, fiberglass sleeves were used (R = 3 per inch). Today, an R-value of 6–7 per inch is achieved with polyurethane sleeves. These are available at most plumbing supply stores and easily pay for themselves in one or two years even though they are more expensive than fiberglass sleeves. Of course any pipes passing through unheated crawl spaces should be insulated.

Clearances. There should obviously be plenty of clear space in front of the boiler for easy loading. In fact, it is a good idea to leave plenty of room on all sides so that maintenance and repairs can be carried out unhampered. The minimum clearance to combustibles recommended by the NFPA is 6 inches on the sides, rear and top.[2] No credit is given for insulation around the boiler. Flow and return pipes should be at least 1 inch from combustibles, except the distance may be reduced to ½ inch where the pipe penetrates a floor, wall or ceiling. Special metal collars available from plumbing supply stores are used to cover pipe holes and prevent infiltration of cold air.

ZONES

It is often desirable to maintain different parts of the house at different temperatures. To some extent this can be done simply by adjusting radiator valves. However, breaking the system up into independently controlled zones gives results closer to the ideal. **Figure 4.4** shows a three-zone one-pipe system in which the main living area, upstairs bedrooms and greenhouse are all maintained at different temperatures. Heat delivery to each zone is controlled by its own thermostat connected to its own pump. Whenever a system is zoned, at least one zone must be available for dumping excess

Figure 4.4 Each zone is controlled by its own thermostat and pump. In one zone—the greenhouse in this case—the overheat (HL) switch must be wired in parallel with the thermostat to provide for dumping excess heat.

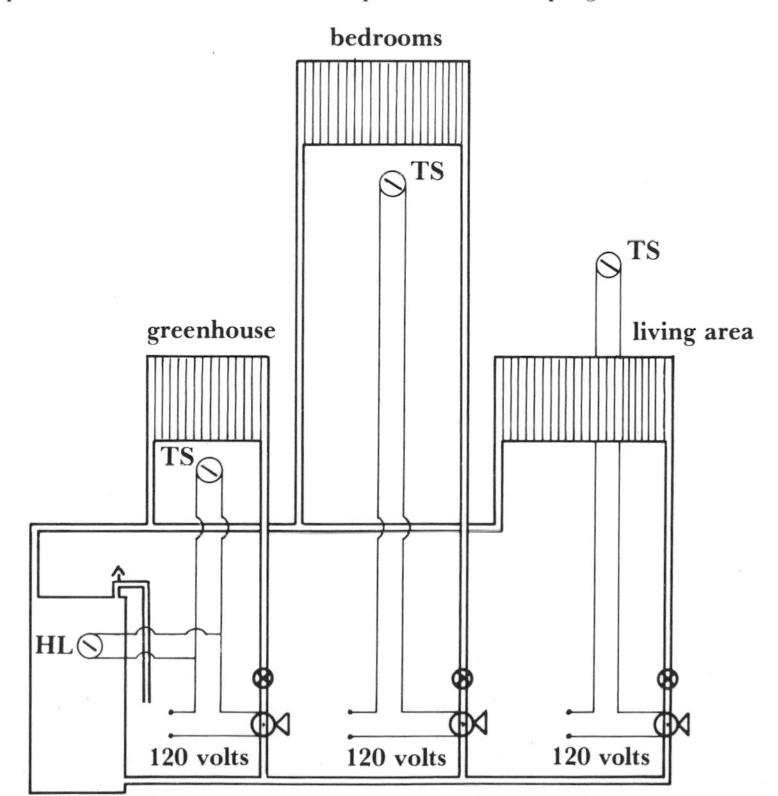

heat. In Figure 4.4 the pump controlling the greenhouse circuit is connected both to the thermostat in the greenhouse and the overheat switch in the boiler.

DUAL INSTALLATIONS

Like furnaces, solid-fuel boilers can be added to oil or gas boilers in series or in parallel. In either case the aquastat of the pre-existing boiler is set lower than the aquastat of the solid-fuel boiler, so that the oil or gas takes over only if the add-on boiler does not maintain the desired boiler temperature. The second boiler enlarges the system to the point that a second expansion tank is required in almost all cases.

Systems in Parallel. The parallel hookup (**Figure 4.5A**) is definitely preferable, in spite of the fact that some manufacturers recommend the series arrangement. When the two boilers are in parallel, water circulates between them constantly, so that the volume of water heated by the solid-fuel side is the sum of the volumes of the two boilers. Conclusion: Overheating occurs less frequently

Figure 4.5A A typical parallel hookup. Pump 2 may or may not run constantly; Pump 1 responds to the thermostat or overheat switch.

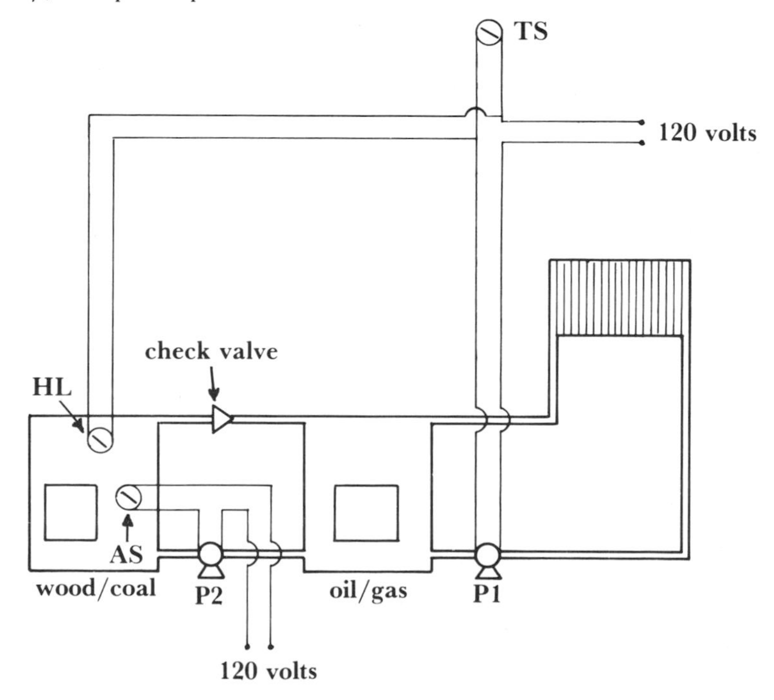

in the parallel system, especially where the gas/oil boiler is large. Unfortunately, many modern oil/gas boilers are on the small side, holding only ten gallons or so.

The interboiler pump in Figure 4.5A is not absolutely necessary, since gravity flow between the two boilers occurs if they are connected as shown. The pump ensures faster circulation and may reduce the likelihood of temperature overshoots by moving heat out of the solid-fuel boiler faster. Plumbers, almost to a man, prefer pumped systems because of their predictability. In view of the fact that most fittings on solid-fuel and oil/gas boilers are too small for good gravity circulation, in almost all cases a pump should be used.

There are two options for controlling the interboiler pump. Either it may be left running constantly, or it may be controlled by an aquastat in the solid-fuel boiler. Choosing between the two depends on how you operate your system. If you rely mainly on solid fuel all winter long, then you might as well let the pump go constantly. This is alleged in fairly well-informed circles to prolong pump life. On the other hand, if you rely on fluid fuel a good bit, then connect the interboiler pump to the power source through an aquastat on the solid-fuel boiler so that the pump goes off when the solid-fuel side is inoperative. This, in combination with a check valve oriented as in Figure 4.5A, prevents flow of hot water from the fluid-fuel to the solid-fuel boiler when the solid-fuel boiler is off. The benefit of this is that it minimizes heat-loss through the solid-fuel boiler, both to the basement and to the flue.

Figures 4.5A and **4.5B** are very much alike except for the inclu-

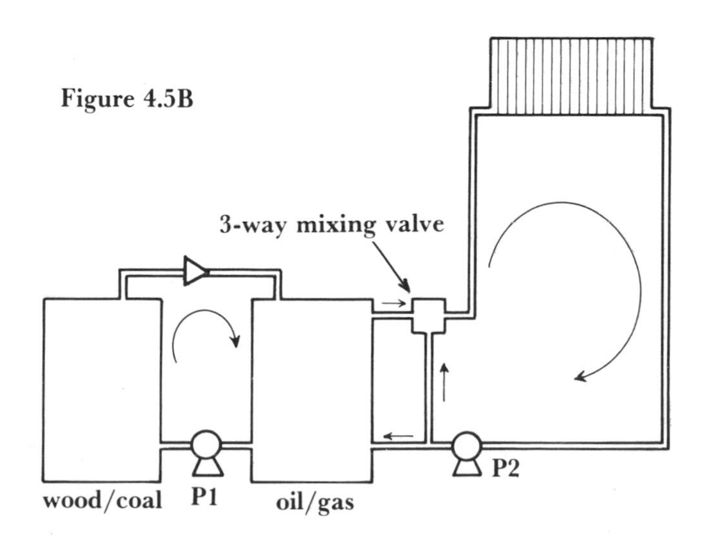

sion of a three-way mixing valve in Figure 4.5B. This mixes return water with hotter flow water in accordance with a command from a proportional thermostat, i.e., a thermostat whose electrical signal is proportional to the need for heat. Both pumps in Figure 4.5B run constantly, not just the interboiler pump. When there is a large need for heat in the house most of the return water goes to the oil/gas boiler. When there isn't, return water is recirculated to the house.

There are several advantages to this constant-circulation system. A big one is that some heat is drawn off the solid-fuel boiler almost all the time, and thus distribution of heat mirrors production of heat in being continuous rather than intermittent. A corollary is that overheating is less likely in a constant-circulation system. Nevertheless, it is good practice to over-ride the mixing valve in case of overheating so that excess heat can be dumped in the house.

Other advantages of the use of a mixing valve in this way are more even heating in the house and longer circulator life, the argument being that the surge of current that accompanies starting a pump motor shortens its life. The mixing valve may be manual instead of automatic, in which case you become the thermostat.

Figure 4.6 shows a variation on the basic parallel theme in which connections are made to the flow and return pipes of the oil/gas boiler instead of to the boiler itself. This is entirely satisfactory and often better than trying to make connections to the original boiler directly, since using a long wrench on corroded plugs may crack the boiler.

Flow to the two zones of Figure 4.6 is controlled by zone valves rather than pumps. This kind of zoning is fairly common because it is cheaper than using a separate pump for each zone. The overheat switch on the solid-fuel boiler should be connected in parallel with the thermostat in the dump zone so that either opens the zone valve and turns on the circulator. This sounds simple, and it really is, but the wires in the basement can be confusing. Shown in Figure 4.6 are two-wire thermostats and zone valves. Another kind of zone valve requires three wires. Don't attempt to do the wiring yourself unless you are sure you know what you are doing. If you make mistakes, you could burn out one or more of the elements in the circuit. It is generally not necessary to alter the mysterious black box that con-

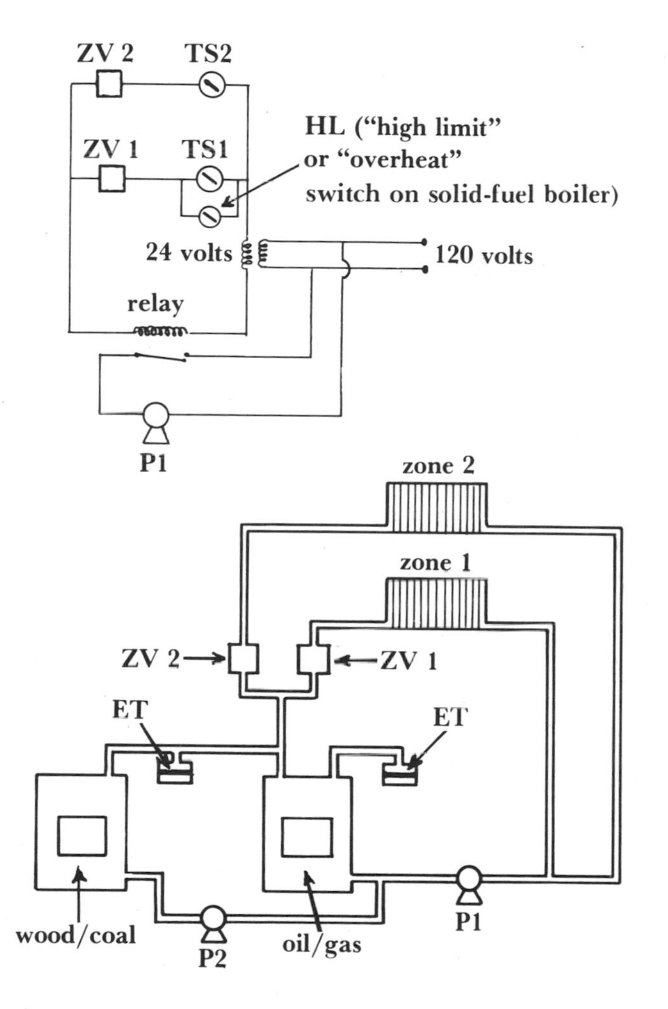

Figure 4.6 When zone valves control flow to the separate zones, the zone valve to the dumping zone must open on a call from the overheat switch, here wired in parallel with the thermostat for zone 1.

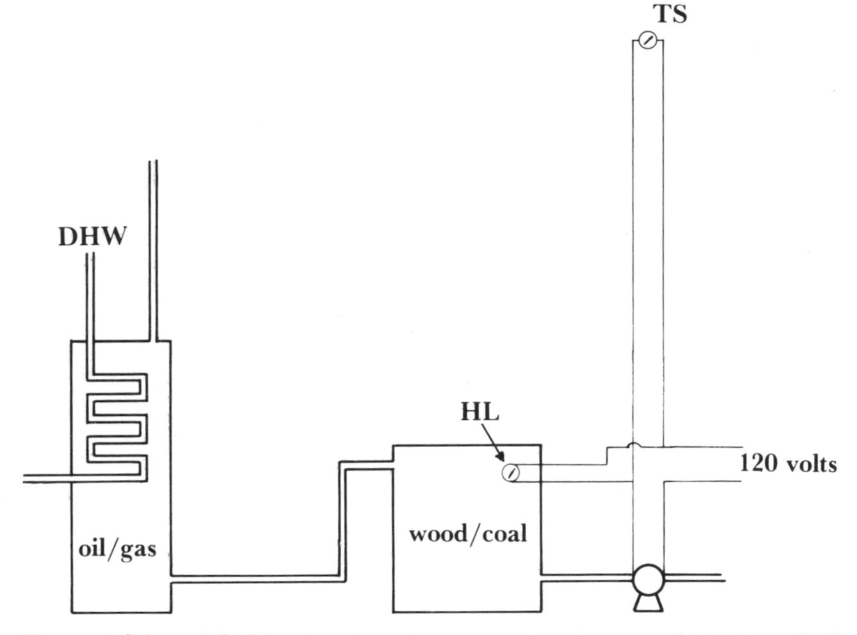

Figure 4.7A and **B** The simple series connection has two liabilities: likelihood of temperature overshoot in the solid-fuel boiler, and cooling of the oil or gas boiler by drawing domestic hot water. The drawbacks are eliminated by incorporating a feedback loop into the system.

trols the gas/oil boiler, except to lower the firing setting from 180°F to 130 or 140°F.

Systems in Series. In the standard series hookup **(Figure 4.7A)** water is stagnant in the two boilers when the upstairs thermostat is not calling for heat. If there is a coil for domestic hot water in the oil/gas boiler, a draw of hot water in the house will cool the boiler and cause the oil or gas burner to fire. In the parallel hookup, however, cooling of the oil/gas boiler in this way can be compensated for by heat from the solid-fuel side. Conclusion: Less oil and gas are burned in the parallel system.

A series hookup can be altered to give it the advantages of the parallel arrangement by introducing a feedback loop. In **Figure 4.7B** an interboiler pump circulates water around an interboiler loop whenever the house pump is off. The controlling element is a relay which senses when there is no electrical current in the circuit including the primary pump. A check valve in the feedback loop prevents return water from bypassing the solid-fuel boiler when the primary pump is on.

BOILER MAINTENANCE

Tube-type boilers must be cleaned from time to time, since creosote or soot deposits reduce heat transfer from exhaust gases to water. Cleaning is done with a wire brush through cleanout doors.

If a small leak should develop, it can usually be taken care of by addition of boiler sealant to the system. This is similar to the stop-leak compounds added to car radiators. Its use should be kept to a minimum because it may clog air vents.

In hard water areas it is advisable to add a corrosion inhibitor to

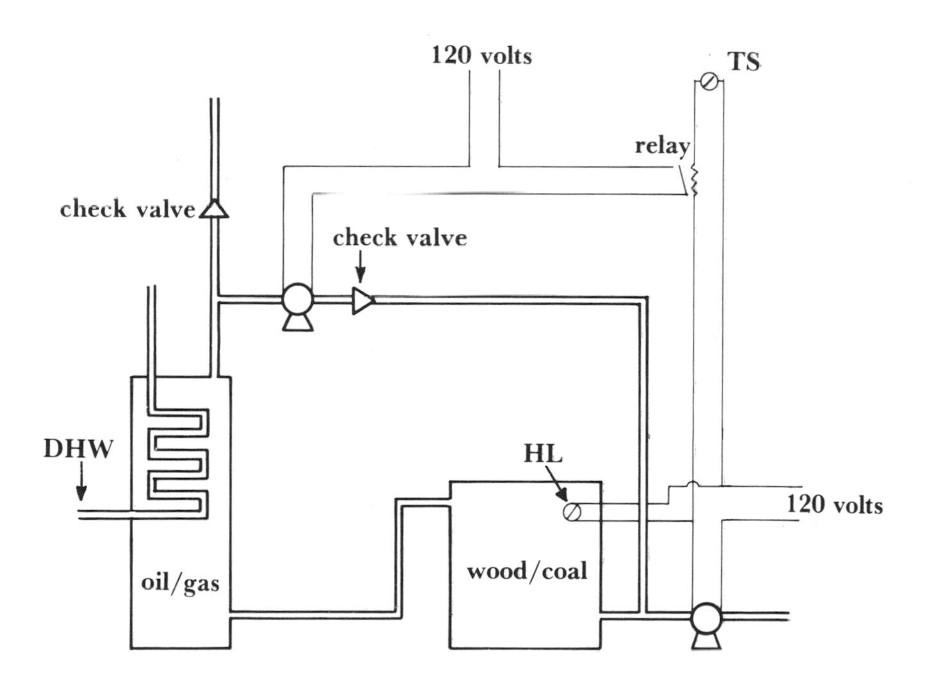

the boiler water. If a boiler is drained for repairs, it should be fired directly after refilling to expel oxygen from the fresh water and thereby prevent it from entering into corrosive chemical reactions. Draining the boiler unnecessarily, in summer for instance, accomplishes nothing except causing more corrosion.

NOTES TO CHAPTER IV

1. "Radiator" is a misnomer in most cases. For radiators operated around 180°F the heat transmitted by radiation amounts to about 40 percent of the total. J. R. Allen, "Heat Losses from Direct Radiation". ASHVE *Transactions,* 1920, page 11.
2. "Heat Producing Appliance Clearances". Bulletin 89M, 1971, National Fire Protection Association, 470 Atlantic Avenue, Boston, Massachusetts 02210.

CHAPTER V

Domestic Hot Water from Solid-Fuel Equipment

Before electricity came to rural America in the late 1930s, farm families got hot water from a storage tank connected to the kitchen range by a thermosyphon loop, that is, by gravity convection, i.e., no pump. The arrangement was usually something like that in **Figure 5.1.** On one side of the firebox there was a heat exchanger variously referred to as a "coil", a "water jacket", "water front", or "water back", and the holding tank was a "range boiler", although boiling in it occurred only rarely, and then not by design. Water flowed into the tank from a spring above the house, and when the tank boiled you could go up to the spring and pretend you were at a spa enjoying the benefits of bathing in the warm springs. The range boiler was generally of steel or, in better families, of copper. It was left uninsulated, since this way the temperature in the kitchen stayed more nearly constant and heat leaking from the tank prevented pipes from freezing overnight. If you wanted a bath in the evening, you had to remember to fire up the stove in the morning. In city apartments domestic hot water came from a steam boiler in winter and from a small auxiliary coal boiler in summer (**Figure 5.2**).

It is not very likely that the old kitchen range will make a big comeback, not now that almost all of rural America is electrified and/or gasified. But there is no doubt that wood- or coal-fired domestic hot water (DHW) can save money today during the winter

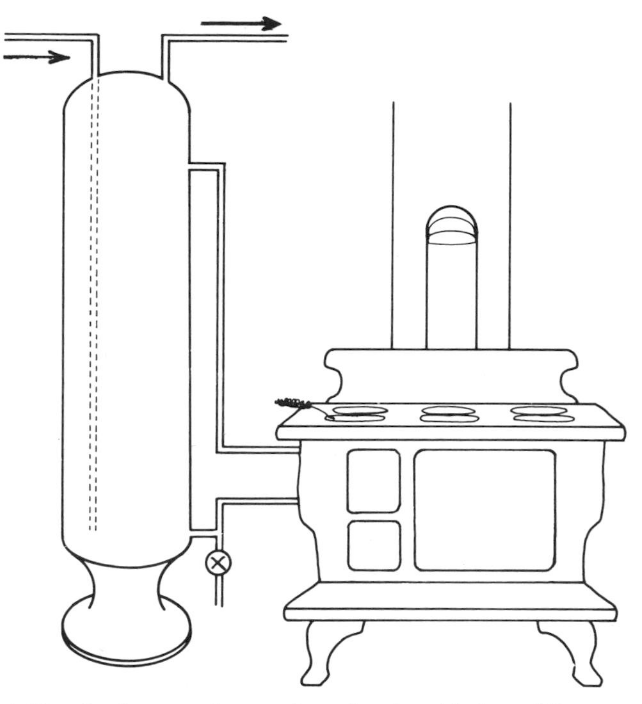

Figure 5.1 Here's the way it was done in the old days. The range boiler stood on a pedestal right next to the kitchen range, which provided food, warmth and hot water. Note the absence of a pressure-temperature relief valve on the tank, implying that the system is open to the atmosphere.

Figure 5.2 In cities many of these small coal boilers can still be found, unused, in dark basements. The balanced lever controls the flow of air into the firebox in accordance with temperature of the water in the flow pipe.

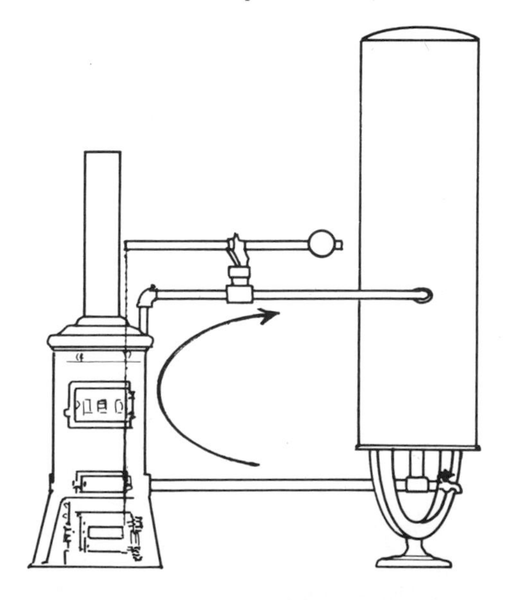

when the stove, furnace, or boiler is going anyway. The sensible course for most people is to use wood or coal for heating domestic water in winter and to use the sun in summer. Payback on a solid-fuel DHW system is generally around three years, whereas for solar it's from six to ten years. By relying on solid fuel in winter, the cost of the solar installation can be significantly reduced, since a solar system intended for summer use only can be smaller and less complex than one intended for year-round production of DHW. The same storage tank can be coupled to both heat sources. The cost of the dual system is less than twice what each one alone would cost. Thus solid-fuel DHW and solar DHW complement one another very nicely.

DHW COILS

Figure 5.3 shows some typical coils, a term covering all these heat exchangers, whether coiled or not. In the old days water jackets were almost always of cast iron, which was thick enough to last a long time even though it rusted. Sometimes coils were constructed

Figure 5.3 Various hot water coils: cast iron, steel pipe, copper tubing.

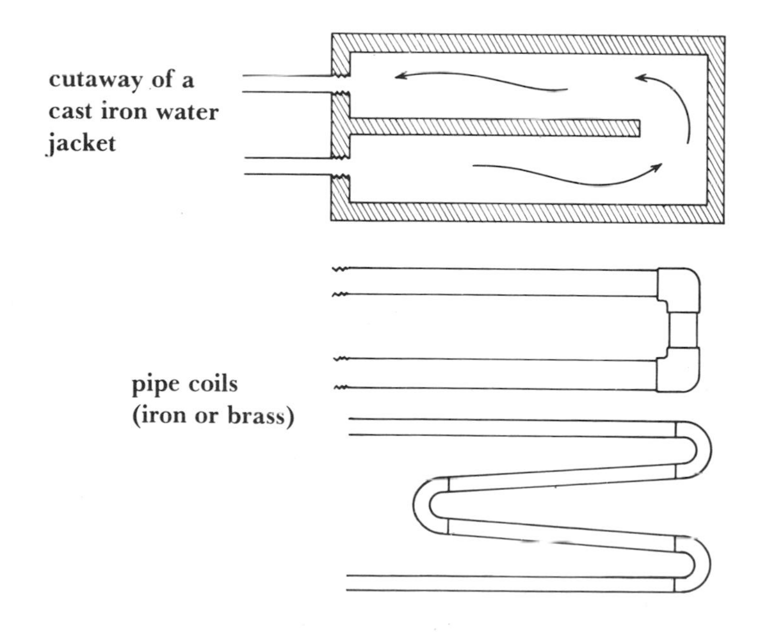

in the field of threaded steel or brass pipe. The latter was more expensive, but needed replacing less often. Today, the Holly Hydro Heater is a prefabricated steel water jacket intended for retrofit on almost any stove or furnace. It comes in various sizes.[1]

Although coils in kitchen ranges were exposed directly to the fire, they need not be. Coils external to the firebox operate at lower temperatures, but this can be compensated for by simply making the coil larger. An external coil is probably safer than an internal coil, and has the nice feature of not taking up space in the firebox. **Figure 5.4** shows one way of adding a coil to a furnace. In this case a roll of ¾-inch copper tubing is simply laid on top of the firebox and hot water rises by gravity into the storage tank above. With a bonnet temperature ranging between 150 and 200°F, it is not too hard to accumulate water hot enough for every domestic purpose. The

Figure 5.4 The coil need not be inside the firebox. If it's external, though, extra length must compensate for lower temperature. Here, the coil is in the bonnet of a gravity hot-air furnace.

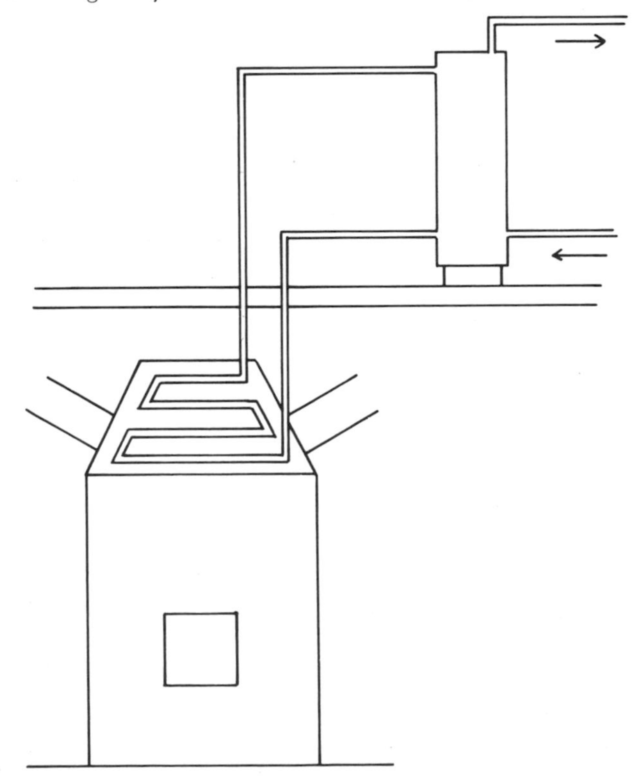

problem is one of control. You must size the coil so that the water in the storage tank is hot enough, but not too hot. As a general design rule, the best approach is to size the coil and the rest of the system so that there is plenty of hot water in the relatively mild months of October and March. This implies water hotter than necessary during January and February when the stove or furnace is going full blast.

There must be provision for converting extra stored heat in the cold months into space heat. This is best done by connecting the storage tank to a radiator in a cold corner or a chilly bathroom and circulating water through the radiator whenever the tank temperature exceeds some limit, 160°F for example. Another, less precise, way to control temperature is to keep the tank insulated during the mild months and remove the insulation during the cold months.

Coils can also be placed in chimneys or stovepipe. Generally speaking, this is not entirely satisfactory, because creosote tends to collect on the coils and decrease their effectiveness as heat exchangers. Only where the flue gases are consistently very hot does a stovepipe heat exchanger perform well. Blazing Showers makes a universal stovepipe heat exchanger by putting 15 feet of coiled copper tubing into a jacket that fits onto the flue collar of almost any stove **(Figure 5.5).**[2]

Figure 5.5 This is the Blazing Showers heat reclaimer. The flue temperature must be consistently high or the coil will become coated with creosote, which seriously interferes with heat transfer to the coil.

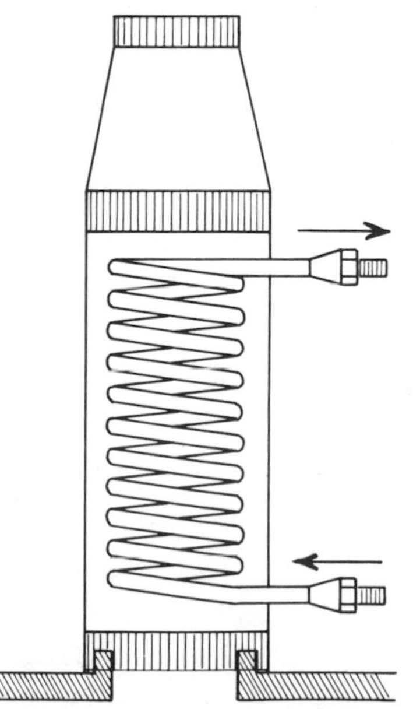

COMPARISON OF PRESSURIZED AND UNPRESSURIZED THERMOSYPHON LOOPS

In **Figures 5.6** and **5.7** the two basic configurations of connecting coil and storage tank are illustrated. The unpressurized thermosyphon loop (Figure 5.6) is to be preferred in areas where ground water is highly mineralized. The reason is that water that passes through the stove or furnace coil is captive in the unpressurized system. Once the minerals have been precipitated at startup, no more scaling can occur in the coil. In contrast, in a pressurized or direct system the water that passes through the coil is the same water that emerges at the tap. If it is highly mineralized, the coil can become occluded in one season. True, the heat exchanger in the tank of the

Figure 5.6 A typical unpressurized thermosyphon. This is also known as the indirect system because tap water does not pass through the stove or furnace coil. The space-heating loop not only provides extra warmth on cold days, but also provides overheat control. If a space-heating loop is not incorporated into the system, a mixing (tempering) valve may be added downstream of the booster to protect anyone at the tap from being scalded. The bypass around the booster allows you to take it out of service when it is not needed.

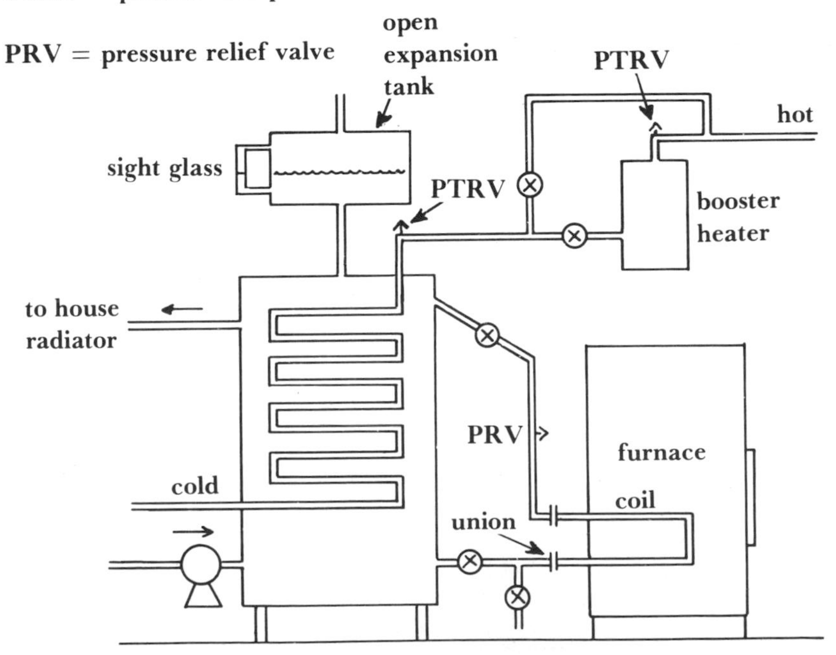

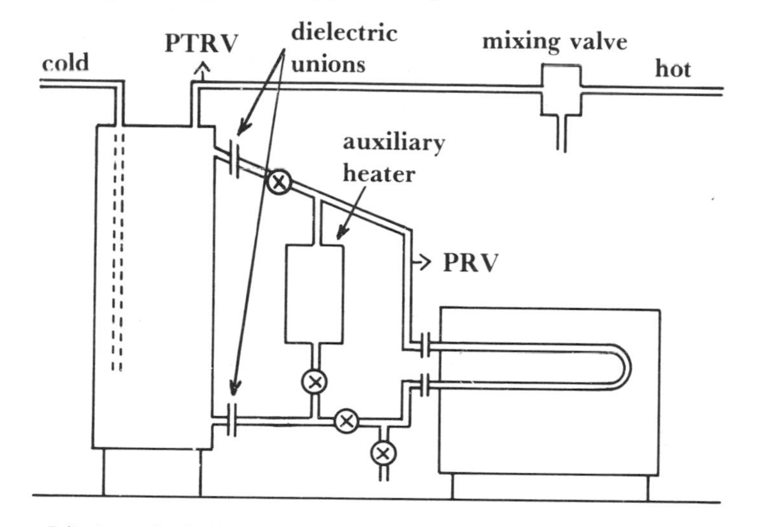

Figure 5.7 A typical direct or pressurized thermosyphon. The gas or electric side-arm heater may be used when the stove is not in operation. The coil is above the bottom of the storage tank so that scale and sediment do not collect in it. If the coil and thermosyphon pipes are copper, insulating (dielectric) unions should be used between the steel tank and copper pipes to minimize corrosion.

One way to provide overheat control is to dump excess heat into a second tank when the primary tank gets too hot. In this case the zone valve opens when the temperature limit is reached, allowing hot water to circulate from the primary to secondary tank and preventing blowoff of the pressure-temperature relief valve.

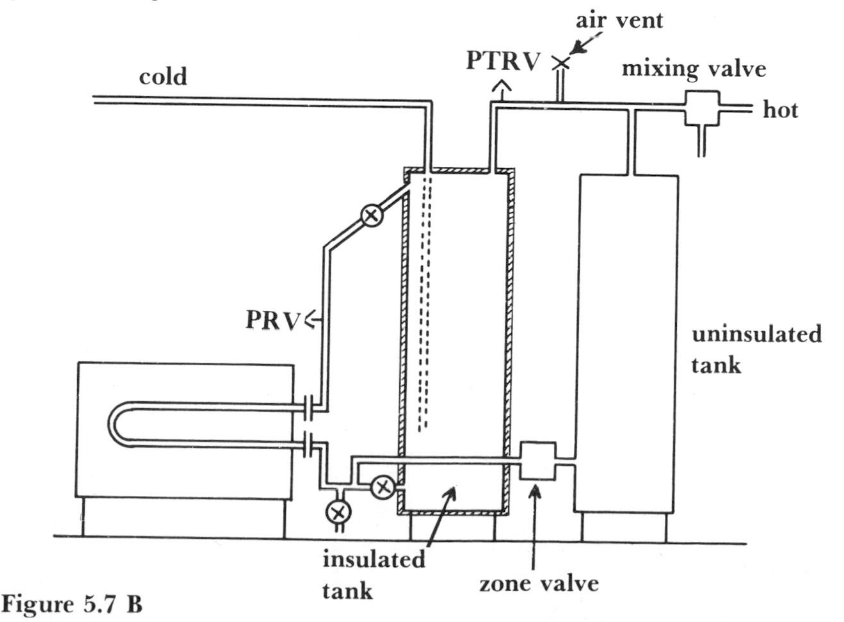

Figure 5.7 B

unpressurized system is subject to scaling, but the problem is less severe than in the coil because the scaling rate is very strongly temperature-dependent. As a rule of thumb, scaling becomes severe above 140°F. The temperature in the stove coil may rise above 200°F in some systems, but the temperature in the tank would generally be less. By dumping heat via a radiator loop whenever the temperature in the storage tank reaches 140°F, scaling in the heat exchanger can be held to a minimum. Note that in the unpressurized system, the radiator loop is also protected against scaling.

Other advantages of the unpressurized system are: (1) The storage tank can be almost anything that will hold water (an old oil drum will do if fittings are welded on in the right places); (2) The unpressurized system is compatible with those solar collectors that cannot stand household water pressure in the range 30–100 psi.

The pressurized system probably has a thermal advantage over the unpressurized system because water at the top of the pressurized tank is not cooled during a draw. Rather, cold water enters at the bottom of the tank and a fairly sharp boundary is maintained between hot and cold water. During a draw this boundary gradually moves upward. On the other hand, in an unpressurized system, as cold water enters the heat exchanger when a faucet is opened, water in the tank in contact with the heat exchanger is cooled. This causes convection currents in the tank which mix hot with cold and keep the temperature in the tank more or less uniform. The consequence is that the temperature falls off rapidly at the tap. This is not disastrous, only a nuisance. It may be remedied through use of a small booster tank in series with the storage tank or through use of an instantaneous heater at the tap.

DETAILS OF A PRESSURIZED THERMOSYPHON

The tank in the pressurized system should be rated for a working pressure of at least 85 psi. This is standard for light-duty range boilers. Gas and electric hot-water heaters are often rated for higher working pressures.[3] The pressure rating of the tank should be stamped right on it. The tank should be lined to prevent corrosion, since the water passing through the tank is oxygenated. Gas and electric heaters are usually lined with glass, which confers a life expectancy on the order of seven to ten years, depending on the mineral content and acidity of the water. "Stone-lined" tanks have life expectancies on the order of twenty years, but cost more initially. The cement lining (about ½-inch thick) impedes corrosion because the rate at which oxygen migrates through the lining to get to the steel is very slow. Corrosion is worst around the fittings, and this is where failure usually occurs. An unlined galvanized steel tank is unsuitable for this application because the protective function of zinc on steel vanishes at about 150°F. This has been attributed to both a reversal in electric polarity between the two metals at this temperature, and also to a change in the physical properties of the corrosion products which form a protective coating at lower temperatures. Whatever the mechanism, no one disputes the occurrence of the phenomenon.

Tank Size. For a family of four a 30- to 50-gallon tank is about the right size for most coils. If the tank is also coupled to a solar collector, the requirements of the solar side of the system should determine the size of the tank, since solar BTUs are harder to capture and more expensive than solid-fuel BTUs. Range boilers tend to be tall and narrow. This favors temperature stratification in the tank, as a result of which a usable quantity of hot water is available at the top of the tank after only a short heating period. After a long heating interval the temperature in the tank becomes more nearly uniform, and when a tap is opened, you can expect to get a volume of hot water about equal to three-quarters of the tank volume. Thus you could expect two fairly long showers from a 40-gallon range boiler and perhaps more, depending on how hot the tank is in the first place. The smaller the tank, the hotter the water and the quicker the recovery. It is not a bad idea to strap an inexpensive thermometer to the top of the tank so you know when the water is ready.

Tank Orientation. Occasionally storage tanks are oriented horizontally rather than vertically. This should be done only where the ceiling is low and it is necessary. Stratification is less pronounced in a horizontal tank, which means that recovery of a usable volume of reasonably hot water takes longer. Another disadvantage is that incoming cold water mixes faster with hot water during a draw.

Expansion Tanks. As with space-heating boilers, allowance for expansion must be incorporated into any pressurized DHW system. In cities expansion may occur into the city water main, but where a check valve prevents this, an expansion tank must be included in the system. In rural water systems the expansion tank is often referred to as the "pressure tank". The formulas of Appendix 2 may be used to determine the correct size for this tank.

Relief Valves. The relief valve on a pressurized DHW system is different from the one on a space-heating boiler in that it is sensitive to temperature as well as pressure. The temperature setting is universally 210°F. The purpose of temperature relief is to protect a person from being scalded at the tap. In a pressurized system the temperature of the water could rise above the normal boiling point of 212°F if there were not temperature relief. When a tap is opened the pressure on the system is suddenly reduced. This introduces the possibility of superheated water (i.e., hotter than 212°F) suddenly flashing to steam and spraying scalding water on the unsuspecting person at the tap. By limiting the temperature in the tank to less than 212°F, this kind of accident is prevented. The relief valve must be installed in such a way that the temperature sensor is immersed in the hottest water.[4]

The pressure setting of the relief valve is normally 150 psi, but this is appropriate only where a gas or electric water heater is used as the storage tank. Rated working pressures of range boilers lie between 85 and 127 psi, and the setting of the relief valve should be chosen accordingly. In those localities where the city water pressure is above the working pressure of the tank, a pressure-reducing valve should be installed on the water line where it enters the house. This valve prevents expansion back into the main, and therefore use of a pressure-reducing valve requires an expansion tank as well.

If the tank temperature is not limited by an automatic heat-dumping loop, it is a good idea to install a mixing valve in the hot outflow line as indicated in Figure 5.7. This cools the hot water to the desired level, usually 140°F, before it reaches the tap. Note: a mixing valve does not supplant the temperature relief part of the temperature-pressure relief valve.

Again, as with space-heating boilers, the pressure-temperature relief valve is rated with respect to the rate at which it can dump heat. It should be big enough to discharge water fast enough to prevent the pressure and temperature of the tank from exceeding relief limits under the worst conceivable conditions. There should be no shutoff valve between the relief valve and the tank, and the discharge pipe from the relief valve should not be obstructed in any way. An important part of maintaining the system is to make sure the pressure relief mechanism is working from time to time by carefully opening the valve. That's what the little lever on top is for. Scale buildup is an enemy of these valves.

Pipes. The pipes in the pressurized thermosyphon loop should be arranged as shown in Figure 5.7. The riser pipe should slope upward toward the tank, so that bubbles of air and gases from dissolved salts can be eliminated from the loop. In a pressurized system there is little chance that air will come out of solution, but it cannot be wholly excluded at high temperatures and low pressures. The lower thermosyphon connection to the tank should be made an inch or two above the bottom to prevent sediment and scale from being carried into the coil.

The Coil. The coil is connected to the rest of the loop through unions, so that it may be disconnected annually for inspection and cleaning. You can't have a very clear idea what's happening in there unless you take it apart and look. The standard method for removing scale from coils is treatment with a solution of hydrochloric (muriatic) acid. Incidentally, dielectric unions, although more expensive, pay for themselves by reducing corrosion at the junction of iron and copper.

To make it possible to disconnect the coil without draining the tank, shutoff valves may be placed in both legs of the loop. If they are, a pressure relief valve must be placed on the coil side of the shutoffs, so that there is no possibility of an explosion should heat be applied to the coil with the valves inadvertently closed. This should be considered an absolute requirement. No exceptions. Incidentally, gate valves are preferable to globe valves because they offer less resistance to flow **(Figure 5.8).**

Three-quarter-inch tubing has become nearly standard for residential thermosyphon loops. This is larger than necessary for heating rates of 5,000 BTU/hr or less when the coil and tank are next to one another, but oversizing the pipe does no harm and is prudent where scaling gradually reduces the size of the passage. Chlorinated polyvinylchloride (PVC) tubing is unsatisfactory in the loop and hot outflow line because its upper working temperature is 180°F. Fifty-fifty tin-lead solder is customarily used with copper tubing in thermosyphon loops, but under extreme conditions its rated working limits may be exceeded. At a temperature of 200°F its max-

Figure 5.8 The advantage of the gate valve is that it offers very little resistance to flow compared to the globe valve. In a thermosyphon the force causing circulation is very weak, and the gate valve is therefore preferable.

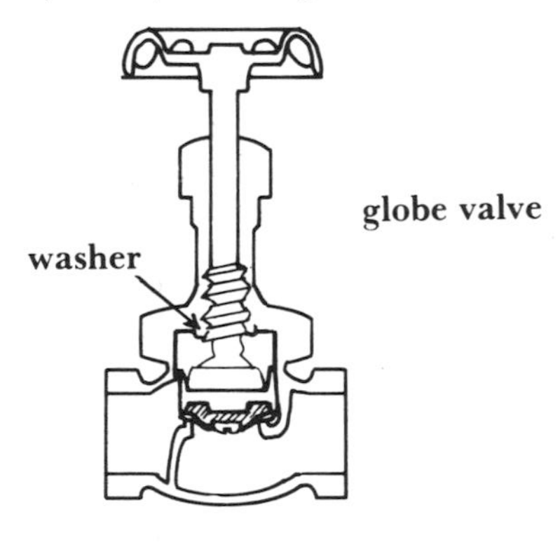

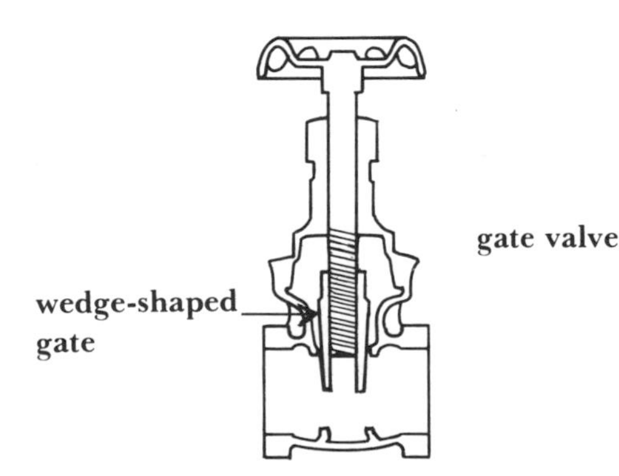

imum working pressure is set at 100 psi. Since the pressure setting of the relief valve is usually 150 psi, at high temperatures a tin-lead joint could become the weakest element in the system. If anything is to blow, you want it to be the relief valve. Therefore, it is good practice to use 95–5 tin-antimony solder, whose working pressure is 300 psi at 200°F.

Tank Position. Heat buildup in the tank is improved very little by elevating the tank. Standard procedure in the old days was to put the tank right next to the heat source and on the same level, as in Figures 5.1 and 5.2. This way the coil was not the lowest part of the loop and sediment stayed in the tank. True, elevating the tank speeds circulation of water in the loop, and this favors heat transfer into the coil (because it keeps water temperature in the coil lower and promotes turbulence), but it also lengthens the loop and thereby increases heat loss from the pipes, which tends to offset the effect of increased flow rate. If you want to put the tank on the floor above to get it out of sight or to take the chill off the bathroom, it certainly won't do any harm, but don't go to the trouble of elevating the tank in the mistaken notion that this will improve hot-water production.

The old rule of thumb, that for every two feet of horizontal distance between the coil and the tank, the tank must be elevated by one foot, seems to have no experimental basis. Clearly there are many thermosyphon systems in operation that violate it. In fact, it is possible to set up a thermosyphon with the coil above the tank as in **Figure 5.9**. A pump is indicated simply because in this geometry transfer of heat to storage would otherwise be slow, but if the power should go off, circulation would continue by gravity and prevent the pressure-temperature relief valve at the coil from venting. When the loop is extended above the coil, hot water in the coil tends to rise to the top of the system, thus driving circulation counter-clockwise. If the loop were not extended above the stove and the power went off, the pressure-temperature relief valve would go off when the stagnant water in the coil reached 210°F. In a rural system where pressure is pump-dependent, loss of water through the pressure-temperature relief valve might drain the coil. Then when power was restored, the overheated coil might be shocked by cool water.

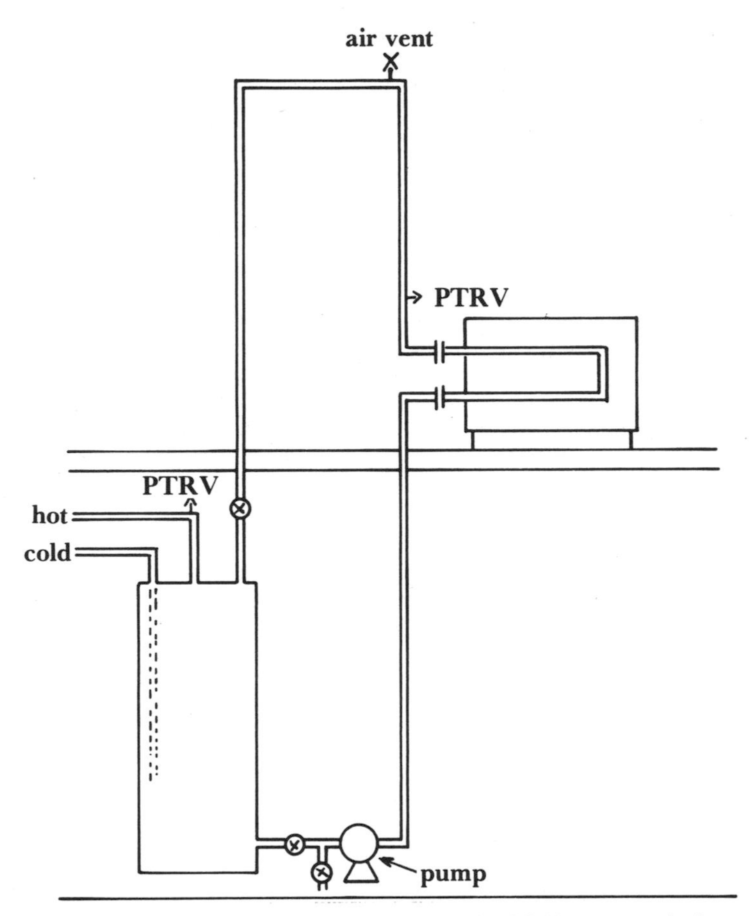

Figure 5.9 It looks strange, but then so did Volkswagens before they became so popular.

This is speculation, since the tests have not been done, but it appears that extending the loop above the coil can confer an element of safety on the system.[5]

THE UNPRESSURIZED THERMOSYPHON

Most of what has been said about pressurized thermosyphons also applies to unpressurized thermosyphons. Circulation in the two systems is unaffected by the difference in pressure between the two, since the driving force, the difference in weight between warmer and cooler columns of water, is nearly unaffected by pressure differences in the range of interest.

Air bubbles will form in the unpressurized system when it is first heated. These are quickly eliminated and cause no trouble if the riser slopes upward toward the tank so they can escape. The temperature of the water in the tank is inherently limited to 212°F, or slightly less at high altitudes, which means that temperature relief on the hot outlet line is not absolutely necessary, although by using a relief valve combining temperature and pressure functions, one reduces the likelihood of violent boiling in the tank.

The storage tank in an unpressurized system may be filled completely and outfitted with an expansion tank of the type shown in Figure 5.6. However, it is easier and cheaper to leave some air at the top of the tank and cover it loosely with a lid so that you can easily disconnect the heat exchanger in the tank for inspection. The tradeoff here is an excessive evaporation rate. The fix is a carefully fitted Styrofoam board that floats on the surface of the hot water. Check the temperature rating of the insulation to make sure you're getting the right stuff.

The heat exchanger in the unpressurized system may be a coil of copper tubing or a small tank immersed in the bigger one. The best choice is probably a finned coil intended for use in a hot-water boiler, since it is designed for adequate heat transfer under similar conditions.

IMPROVING SYSTEM PERFORMANCE

The rate of temperature buildup in the tank of a solid-fuel-fired DHW system depends primarily on the state of the fire, the size of the coil, the size of the tank, and the amount of insulation on the tank. In **Figure 5.10** are plotted the results of a computer simulation of the behavior of a thermosyphon system with varying amounts of insulation on a 40-gallon tank. The first inch of fiberglass makes all the difference in the world. There are diminishing returns with each additional inch. Some foam materials sold for insulating hot water tanks and pipes are not very suitable for use on wood- or coal-fired systems in which water temperature may go above 180°F. The problem is that they may soften and adhere to the copper or steel. If

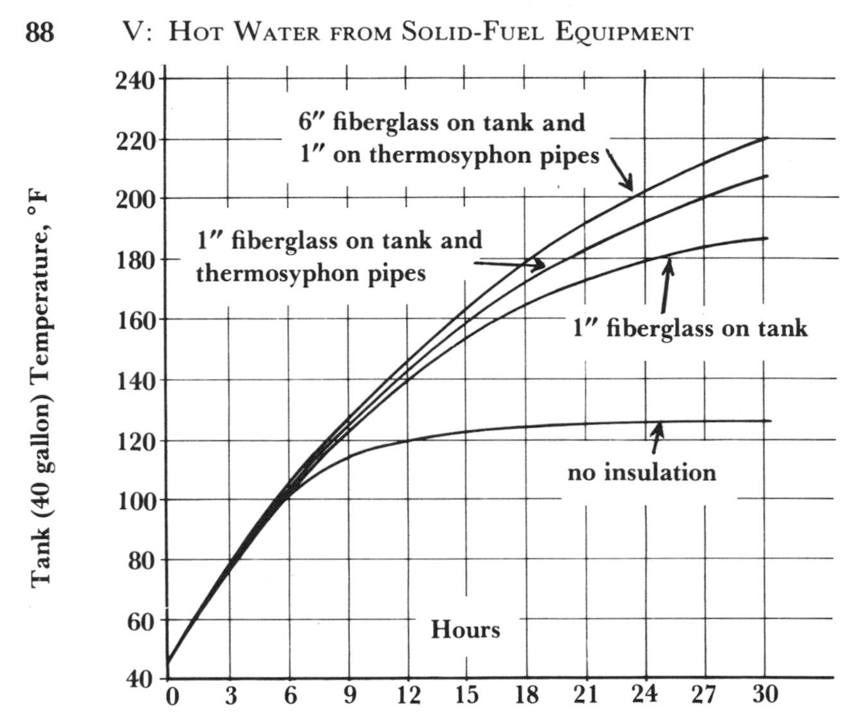

Figure 5.10 The effect of insulation on temperature buildup in a 40-gallon tank connected to a small coil by ¾-inch tubing. Clearly insulating the tank and tubing can make a huge difference in the final tank temperature. The initial heating rate in this case is only about 3000 BTU/hr.

you then want to rearrange the plumbing, you will find that the foam is very difficult to remove and may have to be burned off, creating objectionable fumes. Fiberglass or urethane sleeves rated for at least 250°F are a good choice in this case.

Note that it takes 12 hours to raise the tank temperature from 55 to 120°F, even with 6 inches of fiberglass around the tank. This was with a small coil absorbing only about 3000 BTU/hr. Nevertheless, even with a larger coil hot-water production from a solid-fuel stove or furnace simply is slower than what you are accustomed to. Either you must adjust your schedule to the coil's rate of hot-water production or you must put a conventional water heater in series with the storage tank. This may very well enlarge the system to the point that a second expansion tank is needed.

There are other means of significantly improving transfer of heat to water. One is installing a pump in the thermosyphon line. During cold months when gravity flow provides enough hot water the pump may be turned off. The impeller of a centrifugal pump provides little resistance to gravity flow. But when heat buildup is inadequate, turning on the pump has a marked effect. It introduces turbulence into the water stream in the coil, which speeds heat transfer by bringing cooler water from the center of the pipe into contact with the surface of the coil.

Another tactic that improves heat transfer to an external coil is to improve thermal contact with the stove or furnace by clamping the coil tightly against a hot surface. A final ploy is to insulate the coil on the room side so that it is in a hotter environment. The only suitable insulants for this purpose are vermiculite, perlite, or mineral wool without a resinous binder.

COMBINING SOLID-FUEL AND OTHER DHW SYSTEMS

When a gas or electric heater is used for the storage tank, the situation becomes problematical because the fittings on these tanks are ordinarily not in convenient places. Do not leave the electricity or gas on and expect to save much money with your solid-fuel system. The trouble is that when there is a draw of hot water the electricity or gas will come on and heat the tank up to temperature as though the coil weren't there. In fact, the electricity or gas may overpower the coil when the stove or furnace is relatively cool and cause a reverse thermosyphon, that is, circulation around the loop in the wrong direction with heat flowing from coil to firebox. If this happens often, your electric or gas bill at the end of the month will surprise you by being higher than before you installed the coil. As a compromise you can turn on the conventional heater manually only when you expect to need a large amount of hot water, or you can control the electrical heating element with a timer. It is generally preferable to use a small gas or electric booster tank in series with the storage tank rather than incorporating a backup energy source in the storage tank itself. **Figure 5.11** shows how an electric or gas heater can be used as a storage tank. Water enters through a dip tube whose purpose is to temper the cold incoming water so that it does not shock the tank and crack the glass lining. Since the

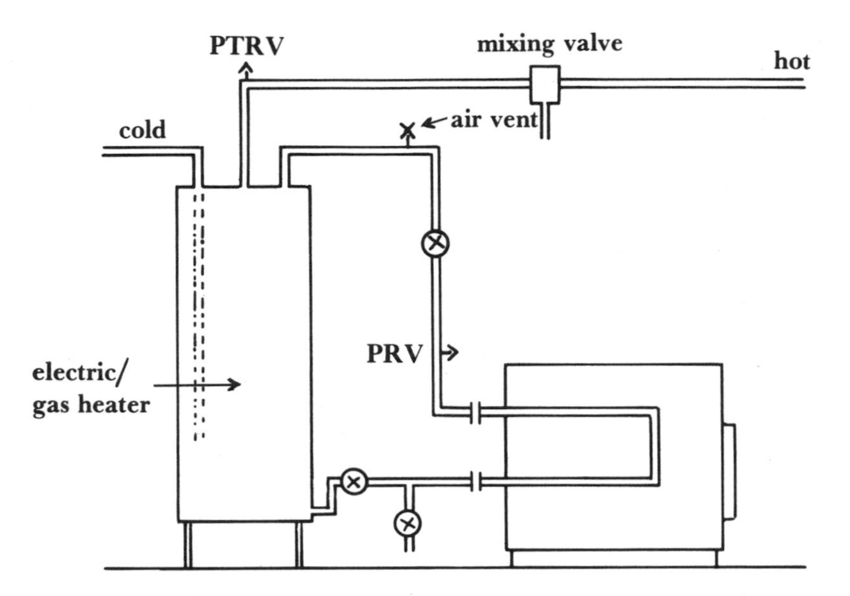

Figure 5.11 The best way of connecting a gas or electric heater and thermosyphon loop if there is no side hole at the top of the tank for the riser. The air vent must be rated for 150 psi, not 30 psi which is common in low-pressure boilers.

tube does not extend all the way to the bottom of the tank, sediment is not stirred up by incoming water. Unless there is a hole on the side at the top of the tank, the riser from the coil must enter the tank through the top. An automatic air vent (rated at 150 psi and not 30 psi) is placed at the high point of the loop to let trapped air escape. There is always a drain fitting at the bottom of the tank where the lower end can be attached. It is always located at the very bottom, precisely so sediment can drain out through it (and into the coil if the coil is below the drain fitting).

If there are only two holes on top of the tank and no side hole at the top, the configuration of **Figure 5.12A** may be adopted. A Venturi tee oriented as shown in **Figure 5.12B** is essential, since otherwise cold water will flow through the coil and enter the hot-water stream at the tee. The Venturi tee prevents this by directing the force of the flowing hot water against the water column in the riser. The configuration of **Figure 5.12C** is not satisfactory because some cold water will bypass the tank through the coil during a draw and

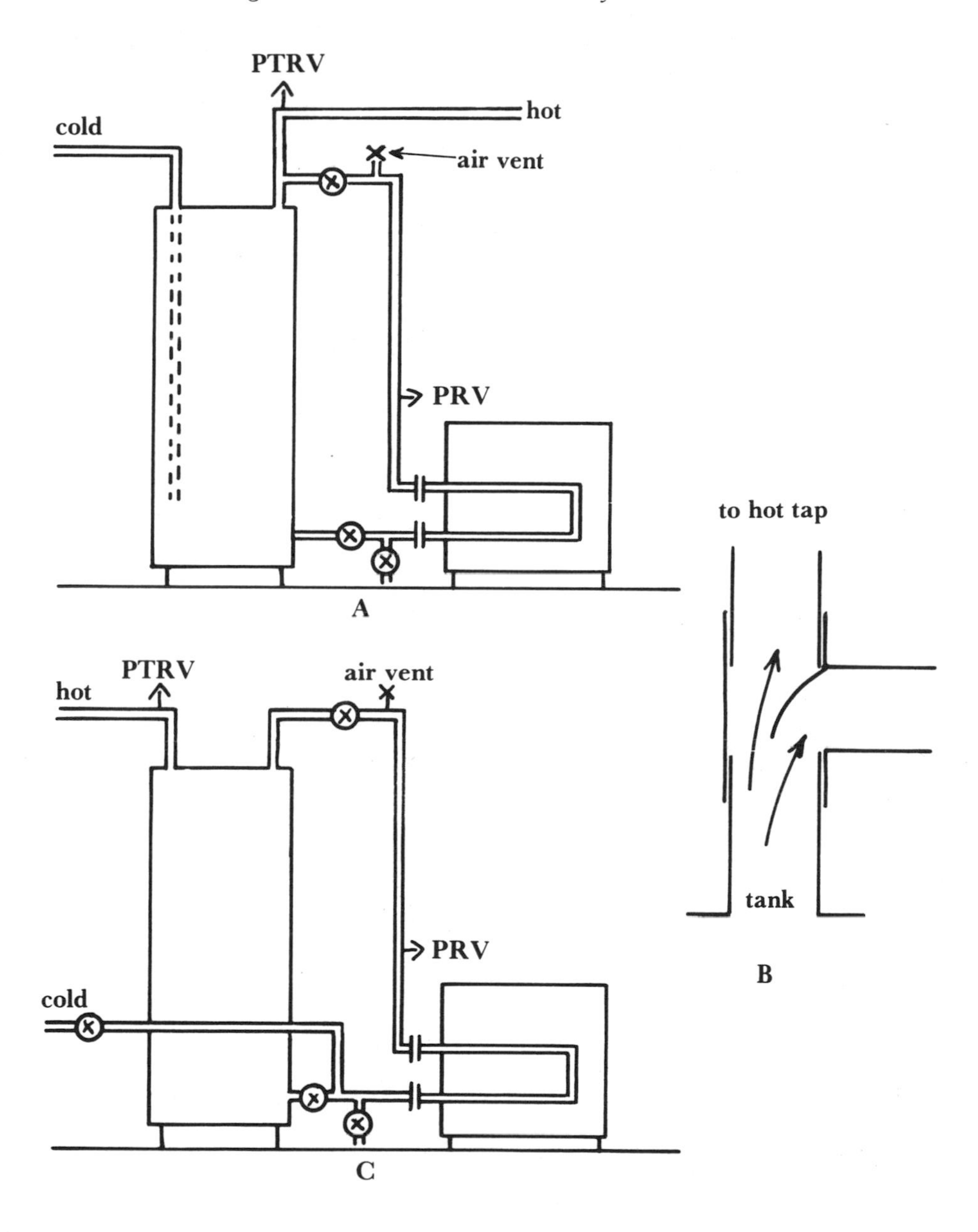

Figure 5.12 When there are only two fittings on top, connection **A** is preferable to connection **C.**

cool hot water at the top of the tank. Furthermore, cold water entering a hot coil may crack it if it is cast iron. Another strike against Figure 5.12C is that cold water enters the tank directly without the benefit of the tempering effect of a dip tube.

NOTES TO CHAPTER V

1. Available from the Frizelle–Enos Company, 265 Petaluma Ave., Sebastopol, California 95472.
2. Blazing Showers, P.O. Box 327, Point Arena, California 95468.
3. Pressure ratings of various storage tanks are confusing. Range boilers are manufactured in accordance with specifications promulgated by the U.S. Department of Commerce in Standard R8–50. There are three classifications of range boiler:

	working pressure	*test pressure*
light duty	85 psi	200 psi
medium duty	106.25	250
heavy duty	127.50	300

   Electric water heaters are rated according to Standard C72.1 of the American Standards Association. This specifies a working pressure rating of 127.5 psi and a test pressure of 300 psi. The same ratings are required for gas water heaters in ASA Standard Z21.10.
4. Strictly speaking, limiting the temperature in the storage tank to 210°F does not in fact forestall superheated water from flashing to steam everywhere. At high altitudes, the boiling point of water may be substantially below 210°F. In Denver, for example, the boiling point of water is 202°F.
5. For further discussion of this, see H. E. Babbitt, *Plumbing*, 3rd ed., McGraw-Hill, 1960, page 206.

APPENDIX I

Furnace-Sizing Method

This method assumes that the maximum heat output rate of the furnace is known with reasonable accuracy and that the balance temperature of the building is 65°F, i.e., that the furnace is needed only at outdoor temperatures below 65°F. For the sake of example assume that previous oil consumption has been 1500 gallons per year and that the number of standard degree-days (i.e., to the base 65°F) is 6000. We also assume the oil furnace is 65 percent efficient, probably a little too generous. Then

Heat loss per degree-day (DD)

$$= \frac{1500 \text{ gal} \times 140{,}000 \text{ BTU/gal} \times .65}{6000 \text{ DD}}$$

$$= 22{,}750 \text{ BTU/DD}$$

Now assume that the lowest outside temperature will be −10°F. Then, on the coldest day, the number of degree-days is 65 − (−10) = 75 DD.

Heat required in 24 hours

$$= 22{,}750 \frac{\text{BTU}}{\text{DD}} \times 75 \text{ DD}$$

$$= 1.71 \text{ million BTU/day}$$

$$= 71{,}000 \text{ BTU/hour}$$

It is prudent to add a safety margin to this. If you install a furnace with an actual capacity of 100,000 BTU/hr you will probably be able to provide enough heat even if the quality of fuel is poor or weather conditions plunge to the one-in-one-hundred-years extreme. This method can seriously underestimate the size of furnace required if the balance temperature is lower than 65°F, as in a well-insulated house, a greenhouse or a restaurant, for example. A correction may be incorporated into the calculations to take care of this effect.

APPENDIX II

Sizing the Expansion Tank

There is a standard formula for sizing closed expansion tanks, namely

$$V_t = \frac{(0.00041t - 0.0466)}{\dfrac{P_a}{P_f} - \dfrac{P_a}{P_o}} \cdot V_s$$

where V_t = minimum tank volume
V_s = system volume
t = design operating temperature, °F
P_a = absolute tank pressure before fill, i.e., pressure relative to a vacuum, not to atmospheric pressure. (Pressure gauges on boilers register the difference in pressure between the boiler water and the atmosphere. To distinguish between gauge and absolute pressures, the abbreviations *psig* and *psia* are commonly used.)
P_f = absolute fill pressure or minimum pressure at the expansion tank
P_o = maximum absolute operating pressure

This formula is a simplified approximation to a more accurate expression; it neglects minor effects such as release of dissolved air from water to the air cushion as heat is applied and the vapor pressure of water in the tank. It is found in both the ASME Boiler Code and ASHRAE *Guides*. The formula assumes that boiling and pressure relief are to be avoided in the normal operating range and that the system is filled with water at 40°F. The numerator gives the expected expansion, taking some credit for the expansion that the pipes and the rest of the system can accommodate as they are heated. Below an operating temperature of 160°F the numerator underestimates expansion and is therefore not valid.

In using the formula, take t equal to at least 180°F to calculate minimum acceptable tank size. If the tank is sized too small, pressure blow-offs may occur. Oversizing the expansion tank does no harm as long as there is an overheat switch on the boiler to keep the temperature below boiling.

For a closed tank without a diaphragm P_a is the pressure of the atmosphere, which is on the order of 14.7 psia at sea level and less at higher elevations, decreasing by about 0.33 percent for every 100 feet in elevation. In Denver atmospheric pressure is about 12 psia.

The fill pressure P_f is the pressure of the air cushion after water has entered the system, but before heat has been applied. It is normally 12 psig, since automatic makeup valves are factory-set for this value. At sea level this makes $P_f = 12 + 14.7 = 26.7$ psia. A fill pressure of 12 psig at the boiler corresponds to about 8 psig at the radiators on the first floor above and about 4 psig on the second floor, because of the difference in hydrostatic head between the basement and the radiators. The object is to keep the pressure in upstairs radiators above atmospheric at all times to aid in eliminating air bubbles from the system.

P_o is the maximum operating pressure, which is set by the relief pressure of the pressure relief valve, 30 psig in most cases. Thus in the typical case, assuming the house is at sea level, we have

$$\frac{V_t}{V_s} = \frac{(0.00041 \times 180) - 0.0466}{\dfrac{14.7}{26.7} - \dfrac{14.7}{44.7}} = 0.123$$

That is, the expansion tank should have a volume no less than 12 percent of the system volume.

The trouble with diaphragmless expansion tanks is that air tends over time to be dissolved in the water and slowly transferred to higher parts of the system where the pressure is lower. The radiators become air-bound and the tank waterlogged. This cannot happen in tanks with rubber diaphragms. Furthermore, they are normally charged with air at the factory to a pressure of 12 psig. Since the automatic makeup valve is normally set for 12 psig, no water enters the tank with diaphragm when the system is initially filled with water. Therefore, the whole tank is available to accommodate expansion.

Going back to the typical case cited above

$$\frac{V_t}{V_s} = \frac{(0.00041 \times 180) - 0.0466}{\dfrac{26.7}{26.7} - \dfrac{26.7}{44.7}} = 0.068$$

Thus the minimum volume of water that a tank with a diaphragm should be able to accommodate is about 7 percent of the system volume.

THE OPEN EXPANSION TANK

In an open expansion tank the entire atmosphere becomes the air cushion. The numerator of the expression on page 95 gives the expansion to be expected when water is heated from 40°F to 180°F. Thus

$$\begin{aligned} \text{Expansion} &= (0.00041 \times 180 - 0.0466)\ V_s \\ &= 0.0272\ V_s \end{aligned}$$

A good rule of thumb is to make the expansion tank twice the size of the expected expansion, in this case 5.44 percent. ASHRAE *Guides* specify 6 percent. Again, it is preferable to oversize the tank rather than undersize it, so that the likelihood of overflow is diminished.

All open expansion tanks should be outfitted with a glass tube so you can keep an eye on the water level. Some evaporation is inevitable, but it can be kept to a minimum by using small-bore vent and overflow pipes. A pigtail in the tube connecting the tank to the rest of the system acts as a heat trap by stopping convection currents which would otherwise carry heat up into the expansion tank. A compromise must be struck between system performance and safety. If the diameter of the tubing is too small, it will prevent adequate pressure relief in the event of boiling.

Although this discussion has been in terms of hydronic heating systems, the same physical considerations apply to domestic hot-water systems. Thus the formulas above may be used unaltered. The major difference is that $P_o = 164.7$ psia in most DHW systems. This decreases the size of the expansion tank required.

APPENDIX III

Design of Thermosyphons

There are standard engineering tables that can be used for determining pipe diameter in gravity flow systems.[1] However, the engineering tables are limited by the assumptions on which they are based and do not lend themselves to modeling the behavior of gravity flow systems over time. Happily, modern computing tools move us a step ahead. Once a calculator or computer is programmed it is literally child's play to determine the importance of various parameters by changing their values one by one and observing the effect on the rate of temperature rise in various parts of the system.

Thermosyphons are out of favor today, the prevailing attitude being to rely on pumps. There is good reason for pumped flow in buildings where the hot water must circulate around large loops. However, in small loops it is better economy to size the pipe large enough for gravity flow and omit the pump, thereby eliminating a source of trouble in advance. Pumps may last for years, but they have been known to fail in less than two.

There are three basic equations to consider:

$$\text{Source heating rate} = W(t_2 - t_1)K \quad (1)$$

$$h_T = h_f \quad (2)$$

$$\left(\frac{dE}{d\theta}\right)_{tank} = C\left(\frac{dt}{d\theta}\right)_{tank} = W(t_2 - t_1)K - \text{losses} \quad (3)$$

Equations (1) and (3) are heat balance equations. The second equation is the steady-state assumption, i.e., that the motive pressure difference—the thermosyphon head h_T—and the frictional pressure difference opposing flow h_f are exactly in balance. This latter assumption is not absolutely correct, but nearly so except at startup.

All three equations must be expanded to be useful. A convenient form of (2) in English units is

$$(\overline{S}_1 - \overline{S}_2) \times h = \frac{f \times L}{64d^5}(5.68 \times 10^{-6}\, W)^2 \qquad (2')$$

$\overline{S}_1$ is the specific gravity of the heavier leg of the thermosyphon loop evaluated at the average temperature of the column. It would be more accurate—but also more cumbersome—to evaluate the integral $\int_0^h (S_1 - S_2)dh$ to find h_T. Equation (2′) is more than good enough for our purposes.[2]

For $Re < 2000$, the straight pipe friction factor, f, is given by $64/Re$. For $Re > 2000$, f can be taken from a Moody chart found in almost any treatise on fluid mechanics. L is the equivalent length of the thermosyphon loop, that is, the length of pipe in the system plus an additional length of straight pipe which would offer the same resistance to flow as the fittings do. There are other ways of handling the resistance of fittings, but (2′) has the advantages of simplicity and utility.[3] Equivalent lengths of various fittings are widely tabulated. There is disagreement in the literature among reported values, which are dependent on details of manufacture, flow velocity, and even position of the fittings in the piping system.

Example 1. A Jetstream boiler (nominal power = 120,000 BTU/hr) is connected to a 500-gallon storage tank as illustrated in **Figure A.1.** Determine the correct pipe diameter for gravity flow. Compare the cost with that of a loop of 1-inch copper tubing and pump of appropriate size.

The first thing to notice about **Figure A.1** is that $h = 4$ ft, i.e., the hot and cold columns causing flow are 4 feet high. True, the overall height of the system is $5\frac{1}{2}$ ft, but the up and down flows above the

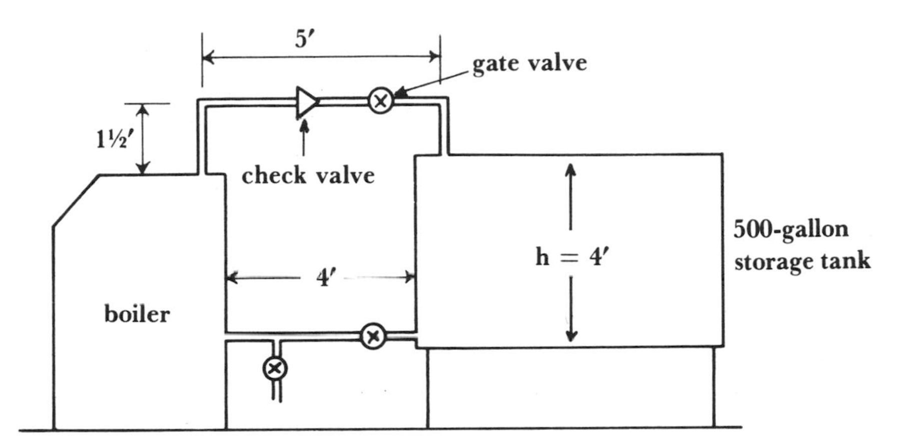

Figure A.1 Thermosyphon Geometry of Example 1

boiler and storage tank are nearly at the same temperature and thus make no contribution to motive force.

In this example we can make some simplifying assumptions just for design purposes.

Assumption 1. That the heating rate of the boiler is 120,000 BTU/hr regardless of the temperature of the water in storage. Heat transfer from combustion gases to boiler water in the 16 tubes of the heat exchanger is probably described fairly well by an equation of the form

$$U_s A_s\left(\bar{t}_s - \frac{t_1 + t_2}{2}\right) = W(t_2 - t_1)K \qquad (1')$$

where the average temperature of the boiler water has been taken as $(t_1 + t_2)/2$. As the temperature of stored hot water rises, the heating rate should gradually fall off. But in our example $\bar{t}_s$ is high, probably on the order of 750 to 900°F, and it may be assumed to dominate the value of the bracketed term on the left in (1′). Thus, for design purposes

$$120{,}000 \cong W(t_2 - t_1) \times 1$$

Assumption 2. That there is no temperature stratification in the storage tank and that a temperature-sensitive switch turns off the combustion blower when $t_1 = 180°F$. Further, we want to keep t_2 below boiling. Let's specify $t_{2,max} = 210°F$. Then, when the storage tank is hottest, (1′) becomes

$$120{,}000 \cong W(210 - 180)$$

and $W = 4000$ lb/hr $= 8$ gal/min.

Now we can use (2′) to determine d. This is not much of a strain with a calculator. If you don't have a calculator, let us hope you remember how to use logarithms and that you are careful and patient. Here are some helpful quadratic equations that are better than 1 percent and 5 percent accurate respectively.

$$S = -1.25 \times 10^{-6}t^2 + 5.83 \times 10^{-5}t + 0.99967$$
$$\nu = 4.61 \times 10^{-10}t^2 - 1.81 \times 10^{-7}t + 2.1221 \times 10^{-5}$$

To solve (2′) for d, it looks as though you just turn it upside down and take a fifth root. The trouble with that is that f and L both depend on d. Our somewhat roundabout approach is to set up a table comparing computed h_f values with the known thermosyphon head as function of various pipe diameters.

L can be found by adding appropriate equivalent lengths. First you must find from **Table A.1** the number of elbow equivalents that the fittings represent. In this case the contributions of 2 elbows (1), 1 check valve (2.5), 2 gate valves (0.2), and 1 tee (0.7) add up to 5.6 elbow equivalents. Then, using **Table A.2**, the equivalent length of the standard elbow is found. Multiplication gives the equivalent length of all the fittings together. To find the overall equivalent length of the circuit, the result is added to the total pipe length, 12 feet in this example. Finally a small contribution for the boiler itself has been added—5 feet for the larger pipe sizes and 7 feet for the 1-inch pipe. The boiler contribution is a semi-educated guess and nothing more.

On the assumption that the copper tubing is smooth inside, f-values have been taken from a Moody chart. Results are summa-

TABLE A.1
Resistances of Various Plumbing Fittings Relative to a Standard 90° Elbow[a]

90° elbow	1
45° elbow	0.5
sweeping 90° elbow	0.7
tee, straight flow	0.7
tee, branch flow	2.0
close return bend	2.4
broad return bend	1.0
open gate valve	0.2
open globe valve	12.0
swing check valve	2.5 and up
union, coupling	0.05

[a] Averages of typical values found in the literature.

rized in **Table A.3.** The calculated friction head for 2½-inch pipe is nearly in agreement with the thermosyphon head of 0.0246 ft. Thus we choose 2½-inch pipe for this hookup, although to play it absolutely safe, the more cautious among us would choose 3-inch pipe. Our calculation has a built-in safety margin for these reasons:

1. The actual heating rate depends on the temperature of water in the boiler and will fall below 120,000 BTU/hr as the system heats up.

TABLE A.2
Equivalent Length of Standard 90° Elbow as a Function of Velocity[a]

pipe diameter	*m*	*b*
½ inch	0.09 sec^{-1}	1.0 ft
1	0.11	2.1
1½	0.17	3.0
2	0.25	4.2
2½	0.32	5.0
3	0.37	6.2

[a] le (ft) $= mu + b$

TABLE A.3
Friction Head as a Function of Pipe Diameter

d	u[a]	l_e for std el	L	Re[b]	f	h_f[c]
1 in	3.27 ft/sec	2.7 ft	34.1 ft	80×10^3	0.019	1.30
2	0.82	4.5	42.1	40×10^3	0.022	0.058
2½	0.52	5.2	46.0	32×10^3	0.023	0.022
3	0.36	6.0	50.6	26×10^3	0.024	0.010

[a] $u = 5.68 \times 10^{-6} W/d^2$
[b] $Re = 5.68 \times 10^{-6} W/\nu d$
[c] $h_T = [S(180°F) - S(195°F)] \times 4 = 0.0246$ ft

2. We have assumed that there is no temperature stratification in the tank at all, which is more or less incorrect. Stratification would mean cooler feed water and hence accommodation of a greater temperature rise in the boiler without boiling. The system is to some extent self-corrective: smaller $d \rightarrow$ lower $W \rightarrow$ stratification $\rightarrow$ accommodation of greater temperature rise in the boiler.
3. We have assumed the existence of a high-limit switch on the storage tank. Its purpose is to stop the flow of combustion air to the fire when the storage tank is up to temperature. If we use 2½-inch pipe and the water in the boiler does boil, we can always lower the high-limit setting.
4. Finally, boiling in an open system of this kind would not be a disaster, only a warning that safety limits were being crowded.

In 1980 the cost advantage lies with the 2½-inch copper pipe and the thermosyphon, assuming that the pump of the alternate forced system would be bronze or stainless steel and correspondingly expensive even though small. It would have to circulate water at a rate of 8 gal/min against a friction head of about 1.3 ft, i.e., it would be a low head pump of fairly high capacity.

Example 2. Suppose a manufacturer sells a furnace with a coil for domestic hot water which he rates at a nominal 5000 BTU/hr. The coil is external to the firebox and feels an average temperature of

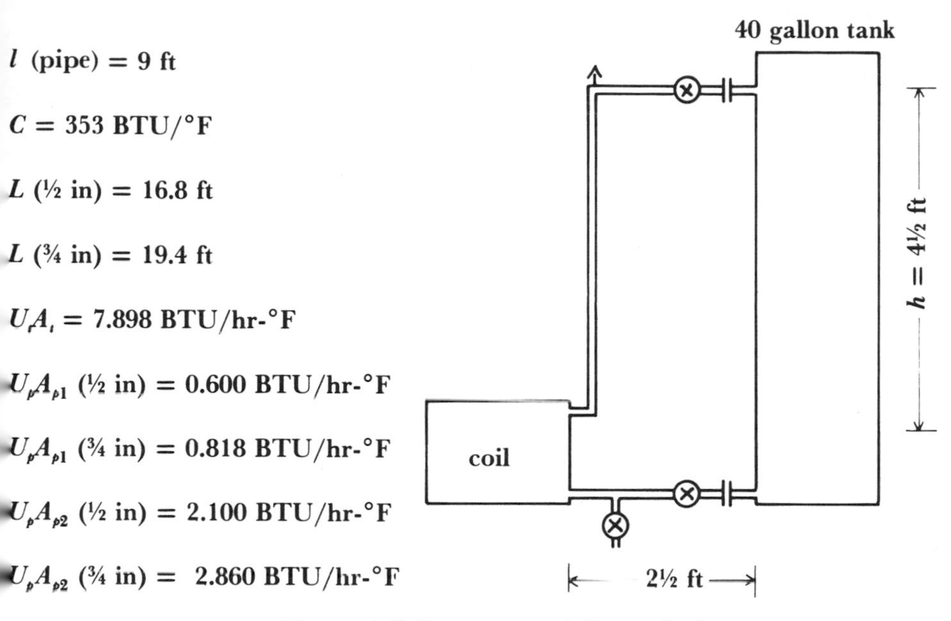

Figure A.2 Parameters of Example 2

300°F. Determine the flow rate, riser temperature and average temperature of the storage tank as a function of time for both ½- and ¾-inch copper tubing. Assume that there is a layer of 1-inch fiberglass on the 40-gallon tank and that the pipes are bare. See **Figure A.2.**

In this problem we cannot assume as we did in the first example that the heat input is independent of tank temperature, since $\bar{t}_s$ is not much greater than the average temperature of water in the coil. In the absence of standard coil rating tests, we simply assume that the manufacturer's rating of 5000 BTU/hr corresponds to fairly cool water in the coil, 60°F perhaps. Then (1′) becomes

$$U_s A_s (300 - 60) = 5000$$
$$U_s A_s \cong 20 \text{ BTU/°F-hr.}$$

Obviously under actual conditions $\bar{t}_s$ may deviate significantly from 300°F and the value adopted for $U_s A_s$ may not be quite right. This does not, however, necessarily invalidate what we are striving for, namely a comparison between two systems under more or less average conditions. If we doubt our results, we can always adopt more extreme values of the various parameters and compare the two systems under extreme conditions. Once the program is in the computer or calculator it's easy as pie to do.

Our strategy is now to solve equations (1′) and (2′) for t_2 and W by successive approximations at successive points in time, starting with $t_1 = 60°$ at $\theta = 0$ hr. Through (3) we determine $dt_1/d\theta$ and hence successive values of t_1 after suitable time intervals. Whether the time interval is short enough can be checked by seeing whether shortening the interval produces significant changes in output. Equation (3) in this example becomes

$$\Delta t_1 \cong \frac{dt_1}{d\theta} \times \Delta\theta$$

$$= \frac{\Delta\theta}{C}[W(t_2 - t_1) \times 1 - U_p A_{p\,1}(t_2 - t_a) - (U_p A_{p\,2} + U_t A_t)(t_1 - t_a)]$$

For small-diameter tubing the equivalent length of the standard elbow is only weakly velocity-dependent. Let us take l_e for the standard elbow to be 1.5 ft for ½-inch tubing and 2.0 for the ¾-inch tubing. We could compute flow velocities and revise l_e values until further interaction produced no significant changes. However, that would simply be a computational exercise without practical effect. Finally, let us guess the resistance of the coil to be 2 relative to the standard elbow.

Note from Figure A.2 that the motive force comes from two columns 5 feet high. We will again assume lack of stratification in the tank. This leads to very little error in $h_T{}^2$. The coil itself has a height of 1 foot. It will simplify matters greatly if we consider the lower half to be filled with water at t_1 and the upper half to be filled with water at t_2. This is equivalent to adopting an effective loop height of 4½ feet.

The results are plotted in **Figures A.3** and **A.4.** After 18 hours the tank is at 180°F for d = ½ inch and 183°F for d = ¾ inch, an insignificant difference. The flow rate is almost doubled by adopting the bigger tubing. This increases heat transfer to water in the coil by keeping t_2 low relative to t_1, but the greater surface area of the ¾-inch tubing negates any gain through an increase in heat loss to the surroundings, which in this case are assumed to be at 70°F.

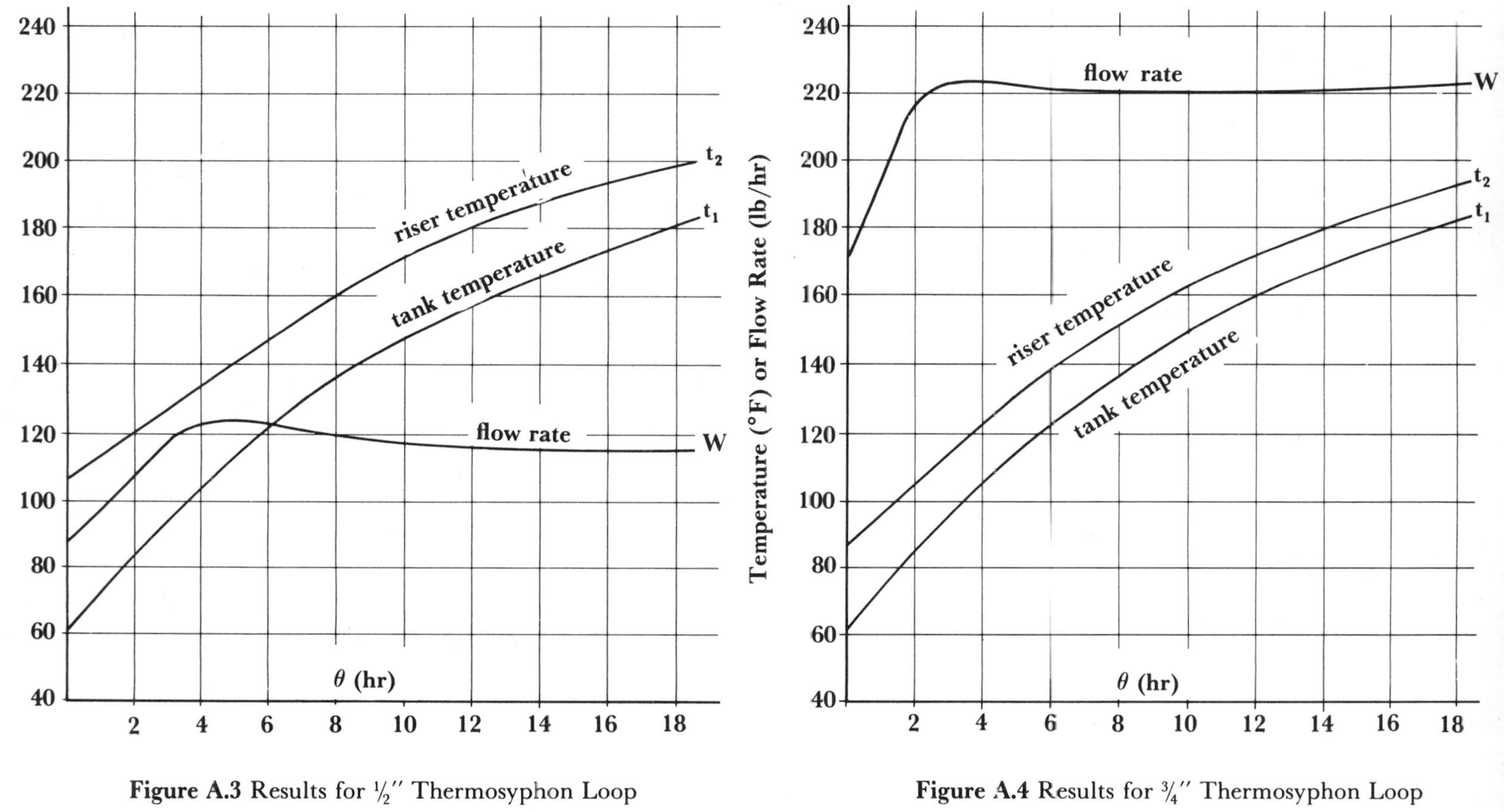

Figure A.3 Results for ½″ Thermosyphon Loop

Figure A.4 Results for ¾″ Thermosyphon Loop

The flow rate abruptly levels off in both cases three or four hours into the run at the transition from laminar to turbulent flow.

In this particular comparison our conclusion is that either ½- or ¾-inch copper tubing would be suitable on the basis of fluid mechanics alone. However, because the smaller tubing would sooner be affected by "crud" or scale buildup, the small extra expense for the larger tubing would be justified.

If the heat source is a solar collector, the same basic equations can be applied with the difference that the left hand side of (1) becomes

$$a \times \sin b\theta - U_c A_c (t_c - t_a)$$

where the first term is the rate of solar input and the second heat loss from the collector to its surroundings.

NOTES TO APPENDIX III

1. In ASHRAE Handbooks and Guides for example.
2. In the second example the exact value of the thermosyphon head is

$$h_T = \int_{1/2}^{5} [S_1 - S_2]\, dh.$$

If we assume a linear temperature distribution in the tank with $t(h = 0') = 120°F$ and $t(h = 5') = 160°F$, then evaluation of the first part of the above integral using the quadratic expression for S given in the body of this Appendix yields

$$\int_{1/2}^{5} S_1\, dh = 0.98261 \times 4.5 \text{ ft},$$

whereas

$$S_1\ (t = 140°F) \times h = 0.98333 \times 4.5 \text{ ft},$$

a difference of only 0.073 percent.
3. The frictional resistance is sometimes written as

$$\left(\frac{fl}{d} + k\right)\frac{u^2}{2gd^2}$$

where k is a resistance term due to fittings. Values of k have been tabulated (see the ASHRAE *Handbook of Fundamentals,* 1977, for example), but depend on flow rates and are highly uncertain. Some are published with generous safety margins incorporated into them.

4. The data underlying Table A.2 emerged from a long series of studies by F. E. Giesecke and students at the University of Texas. See for example, F. E. Giesecke, "The Friction of Water in Iron Pipes and Elbows", *ASHVE Transactions,* 1917, p. 499, and F. E. Giesecke and W. H. Badgett, "Loss of Head in Copper Pipe and Fittings", *ASHVE Transactions,* 1932, p. 529.

NOMENCLATURE FOR APPENDIX III

a amplitude, BTU/hr
b phase factor, hr^{-1}
C heat capacity of tank, BTU/°F
d diameter, ft
f friction factor
h effective height of thermosyphon loop, ft
h_T thermosyphon head, ft
h_f friction head, ft
K heat capacity of water, 1 BTU/lb-°F
l length of tubing, ft
l_e equivalent length of fittings, ft
L sum of equivalent lengths, ft $1 + l_e$
Re Reynolds number
S specific gravity of water
t_1 source inlet temperature, °F
t_2 outlet temperature, °F
t_a ambient temperature, °F
t_c collector temperature, °F
t_s source temperature, °F
u linear velocity, ft/sec
UA heat transfer coefficient, BTU/°F-hr
W flow rate, lb/hr
ν kinematic viscosity, ft^2/sec
θ time, hr

Index